Introduction to Fire Prevention

Sixth Edition

James C. Robertson, MIFireE

Upper Saddle River, New Jersey 07458

Library of Congress Cataloging-in-Publication Data

Robertson, James C. (James Cole)
  Introduction to fire prevention / James C. Robertson. —6th ed.
    p. cm.
  At head of title: Brady.
  Includes bibliographical references and index.
  ISBN 0-13-119031-8
    1. Fire prevention.   I. Title.

TH9145.R55 2005
363.37'7—dc22

2004057281

Publisher: *Julie Levin Alexander*
Publisher's Assistant: *Regina Bruno*
Senior Acquisitions Editor: *Katrin Beacom*
Assistant Editor: *Kierra Kashickey*
Senior Marketing Manager: *Katrin Beacom*
Channel Marketing Manager: *Rachele Strober*
Marketing Coordinator: *Michael Sirinides*
Director of Production and Manufacturing: *Bruce Johnson*
Managing Editor for Production: *Patrick Walsh*
Production Liaison: *Julie Li*
Production Editor: *Karen Fortgang/bookworks*
Manufacturing Manager: *Ilene Sanford*
Manufacturing Buyer: *Pat Brown*
Creative Director: *Cheryl Asherman*
Senior Design Coordinator: *Christopher Weigand*
Cover Designer: *Christopher Weigand*
Cover Photo: *Corbis/Bettmann*
Composition: *The GTS Companies/York, PA Campus*
Printing and Binding: *Courier Westford*
Cover Printer: *Phoenix Color Corporation*

Pearson Education Ltd.
Pearson Education Singapore Pte. Ltd.
Pearson Education Canada, Ltd.
Pearson Education-Japan
Pearson Education Australia Pty. Limited

Pearson Education North Asia Ltd.
Pearson Educación de Mexico, S.A. de C.V.
Pearson Education Malaysia Pte. Ltd.
Pearson Education, Inc., Upper Saddle River, New Jersey

10 9 8 7 6 5 4 3 2 1
ISBN 0-13-119031-8

Dedicated to the Memory of Fire Service and Law Enforcement personnel killed in the line of duty September 11, 2001

Contents

Preface

The prevention of unwanted fires is considered a primary responsibility of the modern fire department, whether career, volunteer, or combination career and volunteer. However, the fire service and others having an interest in fire prevention increasingly recognize that other agencies and organizations, such as health departments, schools, private industries, and civic associations, can be of tremendous help in educating the the public about this important field of personal safety.

It has long been realized that the term *fire prevention* should be broadly interpreted to encompass arson suppression, fire safety education including personal reaction in the event of fire, plan review, inspection, and other elements of code enforcement. All of these areas are examined in this text. In addition, methods to measure the effectiveness of local fire prevention programs are addressed in the sixth edition of *Introduction to Fire Prevention*.

This edition has been updated to reflect current concepts in each of the above-mentioned areas. The major change in building code organizational structure is addressed as is the advent of performance-based codes. The tragic events of September 11, 2001, are noted with comments on lessons learned for code enforcers. Other recent fire tragedies are mentioned as well. Several new evaluation systems for public fire safety education programs are also reviewed. Longstanding research that has ongoing application is retained in the new edition. Chapters have been reorganized to follow a logical sequence and major editorial changes have been made.

ACKNOWLEDGMENTS

The assistance of a number of persons who reviewed and suggested appropriate wording for certain sections is greatly appreciated. These include Cheryl Edwards, David Icove, Lorne MacLean, Robert Neale, L. Charles Smeby, Jr., Clinton H. Smoke, Ashley M. Wood, and John B. Woodall. Many others who contributed updated information and photographs provided invaluable assistance. A complete editorial revision by Kim Borofka was especially helpful. Special thanks to Katrin Beacom and Kierra Kashickey of the publishers and Nancy Marcello, copy editor, for their understanding and assistance.

Thanks also to the following reviewers: Ralph De La Ossa, Long Beach City College, Long Beach, California; Martin Knudsen, M.S., Senior Instructor, Environmental Health and Safety Technology, Texas State Technical College, Waco, Texas; Ernie Misewicz, Fire Marshal, Fairbanks Fire Department,

Fairbanks, Alaska; Ken Riddle, Fire Marshal/Deputy Chief, Las Vegas Fire & Rescue, Las Vegas, Nevada; Clinton Smoke, Asheville-Buncombe Technical Community College, Asheville, North Carolina; and Lloyd H. Stanley, Department Chair, Fire Protection Technology, Guilford Technical Community College, Jamestown, North Carolina.

History and Philosophy of Fire Prevention

1 CHAPTER

◆ **HISTORICAL BACKGROUND**

A great deal can be learned by studying the historical development of fire prevention. It is hoped that the brief review given here will assist the reader in recognizing the reasons for certain procedures being followed in the field today.[1]

As early as 300 B.C., the Romans established a "fire department," composed primarily of slaves. The response of those individuals is reported to have been quite slow. Little else is known about their procedures; however, the program apparently was so unsuccessful that it was necessary to convert the department into a paid force in A.D. 6.

This apparently proved successful, and by A.D. 26 the full-time fire force in Rome had grown to approximately 7,000. These individuals were charged primarily with a responsibility for maintaining fire prevention safeguards. The population of Rome at the time was just under one million.

The fire brigades of Rome patrolled the streets in their efforts to bring about proper fire prevention procedures. They were granted the authority to administer corporal punishment to violators of fire codes and were provided with rods for use in administering such punishment. Records indicate that they were empowered to administer corporal punishment because "most fires are the fault of the inhabitants." It is difficult to imagine public acceptance today of corporal punishment in connection with fire prevention code enforcement. As an interesting sideline, in addition to their fire prevention duties, these patrolmen had the responsibility of keeping a watchful eye on the clothing of individuals who were using the public baths and were required to make inspections of the baths on a regularly scheduled basis to prevent theft.

In 872, according to history, a bell was used in Oxford, England, to signal the time to extinguish all fires. The Norman words *couvre feu,* which meant "cover fire," later developed the English word *curfew.* Records from 1066 indicate that during their occupation of England the Normans strongly enforced the requirements for extinguishing fires at an early hour in the evening. Because construction at the time allowed dwelling fires to easily spread, this preventive measure was an effective safeguard.

As an added means of fire prevention, certain building code requirements were also imposed. Fitzstephen, writing in 1189 during the reign of Henry II, stated,

1

"The only plagues of London were immoderate drinking by idle fellows and often fires.[2] This was an indication of a severe fire problem in that nation in those early days.

In an effort to control the fire situation, the lord mayor of London in 1189 issued an order to the effect that "no house should be built in the city but of stone and they must be covered with slate or tiled." This requirement was apparently vigorously enforced in structures built from that day on in the city. However, the full effect of that order was not yet felt by 1212, when a raging fire occurred in London. This fire took 3,000 lives. No fire before then had ever caused such a great loss of life.

In 1190, Oxford imposed a requirement for fire walls to be placed between every six houses. This was another example of a major city's efforts to rapidly control a fire and limit its spread.

Emphasis was on fire prevention in the Scottish Act of 1426. It ordered, for example, that "no hemp, lint, straw, hay or heather or broom be stored near a fire." Edinburgh merchants selling such wares were permitted to use lanterns, but not candles, and citizens in general were forbidden to carry open flames from house to house.

Fire precautions figured largely in the Edinburgh Improvement Act of 1621. This act ordered noncombustible roofs and required tradespeople who kept "heather, broom, whins and other fuel" in the center of town to remove the material to remote areas.

High-rise buildings were also a problem in 17th-century Edinburgh. Some buildings reached 14 stories, which caused the Scottish Parliament to issue a regulation in 1698 restricting new buildings to a height of 5 stories. The regulation did not, however, affect existing buildings.[3]

History also records problems with arson during riots in the early days of England. In 1272 in Norwich, 34 rioters involved in arson and looting were captured and punished. Their punishment consisted of being dragged about town until dead. One female arsonist was burned alive as punishment for her act.

In a later arson case, the punishment was equally severe. A 15-year-old boy in Edinburgh was judged responsible for setting fire to peat stacks and was burned alive in 1585 as punishment for his act.

Specific punishments for fire prevention violations are also noted in historical documents. The city records of Southhampton, England, contain a late 1500s case in which a baker was fined two shillings for having combustibles too close to an oven. Manchester bakers were forbidden by a 1566 law to keep gorse (barley) "within two bays of the ovens."[4]

Charles II in 1664 gave authority for imprisoning those who contravened building regulations. These regulations related, then as now, to fire safety.[5] A 1763 act prohibited the piercing of fire walls.[6]

Among fire prevention recommendations issued to the public during those early days in England was one in 1643 that suggested that candles be placed in water-base holders. The thought was that an unattended candle would burn down and go out before causing trouble. An act of Parliament in 1556 required bellmen to patrol the streets and cry out, "Take care of your fire and candle."[7]

Fire alarm systems employed by communities in England were unique when compared with present practices. In Nottingham, a patrol of 50 women roamed the streets during the night for the purpose of fire prevention. Their duty, in addition to detecting fires, was to awaken the sleeping community in the event of a fire.

During 1666, a major fire struck London. This fire, referred to as the Great Fire of London, burned for four days and destroyed five-sixths of the city; amazingly only six deaths occurred. This was in sharp contrast with the 3,000 deaths that occurred during the fire of 1212. The effectiveness of the previously imposed fire prevention requirements undoubtedly had a bearing on the reduced number of deaths. Although thousands of structures were destroyed, the progress of the fire was retarded long enough to allow the occupants to vacate their premises.

As a further indication of efforts in the fire prevention field, an English citizen named David Hartley secured a patent for a fire prevention invention in 1722. The invention consisted of steel plates with dry sand between them intended to be used as a means of reducing fire spread from one floor to another. This method of construction was employed in some of the early buildings in England. Mr. Hartley's invention was considered noteworthy enough that a statue was erected in his honor.[8]

In 1794, theater fire protection was given a boost by the placement of a water tank on the roof of a theater in England. The tank provided a curtain of water in the event of a fire. In addition, an iron safety curtain was provided to separate the theater patrons from fire on the stage.[9]

The Birmingham, England, fire brigade issued a requirement in 1884 that inhabited tall buildings be provided with two staircases. This was looked on as a progressive fire protection measure.[10]

◆ EARLY FIRE PREVENTION MEASURES IN NORTH AMERICA

Massachusetts Governor Winthrop, as a result of a serious fire in Boston, issued an order in 1631 prohibiting building chimneys of wood and covering houses with thatch.[11]

Fire inspections in the New World probably began in 1648 when Governor Peter Stuyvesant appointed four fire wardens to inspect wooden chimneys of thatched roof houses in New Amsterdam (New York City). These individuals were empowered to impose fines for improperly swept chimneys.[12]

Salem, Massachusetts, in 1644 imposed a fire safety ordinance requiring inhabitants to procure ladders for their houses. This was followed by a 1663 ordinance requiring that chimneys be swept each year.[13]

Chimneys were a major fire problem in Colonial America. Philadelphia found it necessary in 1696 to prohibit the cleaning of chimneys by burning them out. Citizens were not allowed to smoke on the street at any time, and the possession of more than six pounds of gunpowder within "forty paces of any building or dwelling" was prohibited.[14]

Norfolk, Virginia, prohibited wooden chimneys in 1731.[15] Easton, Maryland, in a 1791 ordinance, required chimneys to be built of brick or stone.[16] In 1796, New Orleans, then a Spanish province, passed an ordinance against the use of wood roofs.[17]

Rhode Island's first fire prevention law was enacted in 1704. It banned the setting of fire "in the woods in any part of this colony on any time of the year, save between the tenth of March and the tenth of May annually nor on the first or seventh day of any week." A subsequent measure enacted in 1731 prohibited unauthorized bonfires.[18]

Fire prevention enforcement measures were initiated in many communities during the early days of our country. As an example, in 1785 a city ordinance in Reading, Pennsylvania, imposed a fine of 15 shillings for each chimney fire that occurred in the city.[19] The fine was collected by the city and turned over to the fire company that had responded to the alarm. This ordinance was later repealed. Another Reading requirement was the alteration of chimneys in blacksmiths' shops to make them fire resistive, with a fine of $20 for violation.[20]

An 1807 ordinance in Reading prohibited the smoking of cigars on the street after sunset.[21] It also prohibited people from sitting on porches or in the doorway of any house with a lighted cigar or pipe without the consent of the owner. A $1 fine was imposed for violations of this ordinance. Under the Reading ordinance, the use of fire-crackers was prohibited, with a fine of $1 or 12 hours in jail for violators. A duty was imposed on the citizens of Reading to confiscate and destroy fireworks found in the possession of a child.

The Board of Aldermen in Pensacola, Florida, passed an ordinance in 1821 requiring chimneys to be kept swept. A $10 fine was levied against the owner of any house whose roof caught fire.[22]

Jamestown, New York, imposed fire prevention regulations in 1827. Fire wardens were required to examine all chimneys, stoves, and other fireplaces used within Jamestown to direct "such reasonable repairs, cleansings, removals or alterations as shall be in his or their opinion best calculated to guard against injury by fire." Fines were imposed for failure to comply or for refusal of entry to the warden.

Occupants of shops or other places in Jamestown where rubbish might accumulate were required to remove accumulations as often as the warden saw fit. Fines were imposed for each day the violation continued.[23]

The first fire safety ordinance in Greensboro, North Carolina, enacted in 1833, required each household to have two ladders on its premises, "one which shall reach from the ground to the eaves of the house, the other to rest on top of the house, to reach from the comb to the eaves." Two inspectors were appointed to enforce this requirement and to ensure that all rubbish and nuisances were cleared from backyards. A five-dollar fine was imposed for each violation.[24]

In most newly formed towns, fire suppression forces were organized before the advent of fire prevention efforts. However, in Auraria, the original section of what is now Denver, Colorado, the legislative council in 1860 appointed six fire wardens "to inspect buildings and their chimneys and to prevent the accumulation of rubbish" as the result of a large livery barn fire. The first fire fighting company was formed in 1866.[25]

More comprehensive fire prevention regulations were imposed in New York City in 1860 subsequent to a tenement building fire in which 20 people were killed. This ordinance required all residential buildings built for more than eight families to be equipped with fireproof stairs and fire escapes.[26]

Several major fires occurred in the early 1800s in Montpelier, Vermont. As a result of these fires, "the village appointed a committee of three to report a code of by-laws for the preservation of buildings from fire. The bailiffs were required to inspect every house in their ward to see that there was no fire hazard and that each place had, as the by-laws required, a fire bucket and ladder." Another by-law required that no fire should be left burning in a house unoccupied between the hours of 11:00 P.M. and 4:00 A.M., if adjacent to another.[27]

Fire escapes and exits attracted the attention of the Boise, Idaho, city council in 1887, when they imposed a requirement that doors on halls in theaters be made to swing outward. They were concerned about the possibility of a disaster at a performance in one of the city's places of assembly.[28]

Fire alarms and fire escapes, of course, had been invented, but they were not yet generally accepted. In fact, the Illinois legislature attempted to enact a fire escape law in 1897, one which would have replaced earlier, ineffective legislation. The 1897 act required fire escapes in all buildings more than four stories high and in all buildings higher than two stories if these structures were used as manufacturing places, hotels, dormitories, schools, or asylums. According to the *Centennial History of Illinois*, this act was bitterly fought by the Manufacturers' Association of Illinois. When passed, it proved impossible to enforce, and it was repealed in 1899. As late as 1912, 308 fire deaths were reported in Illinois, slightly less than half of these occurring in Cook

County alone. Most victims were trapped in burning buildings. This suggests a continuing problem with fire exits and escapes, although the circumstances of the deaths were not individually reported.[29]

An 1896 fire that destroyed a saloon and hotel brought about the first fire prevention code in West Palm Beach, Florida. The ordinance established a fire district in which no building could be erected unless it was of brick, brick veneer, or stone construction.[30]

Captains of steam fire engine companies in Memphis, Tennessee, as early as 1900 performed inspections to locate and correct rubbish conditions in buildings, dangerous stove pipes, obstructed fire escapes, and defective chimneys and flues. Great amounts of cotton were stored on vacant lots and on streets, which further contributed to the fire problem.[31]

Formal fire prevention measures in Tulsa, Oklahoma, apparently began with a 1906 requirement that owners of all buildings with three or more stories install fire escapes. Failure to comply by a set date resulted in a fine of $15 per day. Store owners were prohibited from using rubber tubing for gas connections. Failure to comply resulted in steel piping being installed at the expense of the owner.[32]

Fire chiefs in the United States have long had an interest in fire prevention. The first general topic discussed at the First Annual Conference of the National Association of Fire Engineers (predecessor to the International Association of Fire Chiefs) was fire prevention. The following specific fire prevention items were discussed at the conference, which was held in Baltimore in 1873:

I. The limitation or disuse of combustible materiel in the structure of buildings; the reduction of excessive height in buildings and the restriction of the dangers of elevator passages, hatchways, and mansards.

II. The isolation of each apartment in a building from other apartments, and of every building from those adjoining by high party-walls.

III. The safe construction of heating apparatus.

IV. The presence and care of trustworthy watchmen in warehouses, factories, and theatres, especially during the night.

V. The regulation of the storage of inflammable materiel, and the use of the same for heating or illumination; also exclusion of rubbish liable to spontaneous ignition.

VI. The most available measures for the repression of incendiarism.

VII. A system of minute and impartial inspection after the occurrence of every fire, and rigid inquiry into the causes, with reference to their future avoidance.

VIII. Fire escapes actually serviceable for invalids, women, and children.[33]

The use of fire suppression personnel for prefire planning inspections was discussed by the Salt Lake City fire chief at the 1901 conference. At their conference in 1902, fire chiefs discussed developments in fire-retardant paint and slow-burning wood.[34]

In Milwaukee, fire prevention requirements were first imposed for places of public assembly in 1888. Violations of these regulations carried fines of $5 to $100. Apparently, these were the only regulations of a fire prevention nature in effect in the community at that time. By 1913, Milwaukee had a force of 30 men devoted entirely to fire prevention duties. There were 90,000 buildings in the city at that time. The personnel assigned to fire prevention duties were paid entirely through the returns from an insurance premium tax. By 1919, more than 250,000 inspections were being conducted each year by this fire prevention bureau.[35]

Fire prevention bureaus were started after 1900 in a number of larger cities. Long Beach, California, established such a bureau in 1917, while Phoenix, Arizona, started its in 1935. At that time Phoenix had a population of 46,500.[36,37]

The development of water distribution systems has played a major role in community fire defense. In Houston, Texas, the first fire engine arrived in 1839; however, a public waterworks did not come about for many years. By the mid-1870s, most businesses had cisterns for fire protection. In late 1878 the city signed a contract for the development of a water distribution system, which was in service by the following summer. This pattern of water system development is typical of North American cities.[38]

Unfortunately, some of these fire safety provisions were not effective, as noted in the following report from Evansville, Indiana:

> As time passed without a big fire, the city grew lax. In spite of the ordinance against frame buildings within the fire limits, the Council routinely allowed variances. Other builders simply violated the building codes. The tightly packed frame buildings were rightly perceived as a fire hazard. In June 1850 the Council required the city marshal to begin investigating all building code violations within the fire limits. They also asked the city attorney to determine whether they could prosecute carpenters, brick and stone masons and "other mechanics" who violated the codes.[39]

◆ TRAGEDY A SPUR TO REGULATIONS

It has been said that "in the realm of fire 'the law' is a thing mothered by necessity and sired by great tragedy."[40] The truth of this becomes clearer when reviewing some of the major fires that have occurred through the years in context with the development of fire safety regulations and procedures in the United States.

FIGURE 1.1 ◆ Inspection of historic structures such as the Thomas Wolfe home site is an important safety measure. *(Asheville, North Carolina, Fire Department)*

PUBLIC ASSEMBLY

On December 5, 1876, a major fire occurred in the Brooklyn Theater in New York. In this fire a stage backdrop was ignited, and 295 people were killed under conditions similar to those in the Iroquois fire 27 years later.[41]

The Iroquois Theater fire in Chicago, outstanding among public assembly fires, occurred in 1903 during a Saturday matinee of a new play, *Mr. Bluebeard.* The Iroquois was Chicago's newest theater and was also considered its safest; it was advertised as being "Absolutely Fireproof." There were 2,000 people present for this performance. Arc lamps were used in the theater. A light set a curtain on fire, and flames and smoke rapidly made the structure untenable. Despite heroic efforts, panic ensued, and human logjams developed at each of the doors. No fire extinguishers were provided. The curtains were combustible, and exits were improperly marked and swung inward. No venting was provided for the stage area, and there was no means of immediately removing hot gases and smoke. This tragic fire took 603 lives and provided a great impetus to the fire prevention movement, especially in the field of public assembly occupancies.

In Natchez, Mississippi, a fire in a small dance hall, the Rhythm Club, took 207 lives and caused injuries to 200 more in 1940. Combustible decorations and one exit with the door opening inward were the factors responsible for this tragedy. Over 700 had been packed into this one-story building, which measured only 120 feet by 38 feet.

During the early days of World War II, a major fire struck the Coconut Grove nightclub in Boston, Massachusetts. On the night of the fire, November 28, 1942, the club had approximately 1,000 occupants, many of whom were people preparing to go overseas on military duty. A lighted match used by an employee changing a lightbulb has been considered the possible cause for this tragic fire, which took 492 lives. Almost half of the occupants were killed, and many were seriously injured. Flammable decorations spread the fire rapidly. Men and women were reported to have clawed inhumanly in an effort to get out of the building. The two revolving doors at the main entrance had bodies stacked four and five deep after the fire was brought under control. Authorities estimated that possibly 300 of those killed could have been saved had the doors swung outward. It should be noted that the capacity of the structure had also been exceeded.

The Coconut Grove fire prompted major efforts in the field of fire prevention and control for nightclubs and other related places of assembly. Immediate steps were taken to provide for emergency lighting and occupant capacity placards in places of assembly. Exit lights were also required as a result of the concern generated by this fire.

Fire protection under the big top was emphasized by the fire that struck the Ringling Brothers, Barnum and Bailey Circus while playing in Hartford, Connecticut, on July 6, 1944. Seven thousand people attended the daytime performance. The circus tent, which measured 425 feet by 180 feet, was apparently not properly flame-retardant, and the fire caused 163 deaths and 261 injuries. After this fire, many states and municipalities gave more attention to circus fire safety requirements. It is ironic that the fire occurred in Hartford, a city that had had an outstanding fire prevention program for many years.

On May 28, 1977, a tragic fire struck the Beverly Hills Supper Club in Southgate, Kentucky. At the time of this fire, which took 164 lives, the club was occupied by 3,000 to 3,400 people. The building, which had an area of 54,000 square feet, was of unprotected, noncombustible construction. Fire separations, automatic sprinklers, and other safeguards were lacking. Exits were insufficient for the capacity crowd. Interior furnishings were made of combustible materials.

The Beverly Hills fire spurred new demands for improved fire safety measures, including inspection improvements. Many of the patrons in the club at the time of the fire were from other jurisdictions that strongly enforced codes for public assembly occupancies. National political leaders raised the question of the propriety of citizens of one jurisdiction being exposed to fire danger when visiting an area where code enforcement is not as stringent. The impact of this fire would be felt for many years to come.

On February 17, 2003, Chicago was again in the national news when 21 persons were killed and 57 injured attempting to leave a night club where mace had been used to quell a fight. Three days later 100 died in a West Warwick, Rhode Island, nightclub fire. Code impacts are noted in a later chapter.

VESSELS

Another example of a fire which had a major impact on a fairly limited occupancy arena was the disaster that struck the excursion steamer, *General Slocum,* on June 15, 1904. This vessel, which had been constructed primarily of wood, steamed down the East River in New York with 1,400 passengers on board. Within half an hour, fire was discovered on the forward deck. Efforts by the untrained crew to control the fire were futile because the hose burst upon being pressurized. Life preservers were faulty, and lifeboats were entirely inadequate. The vessel was eventually beached, only to have had 1,030 persons perish either from the effects of the fire or by drowning.

This tragedy led President Theodore Roosevelt to appoint a commission to study the disaster and to make recommendations for future action. The investigation found officers of the Steamboat Inspection Service to have been negligent in their duties. The president ordered the dismissal of these individuals, and Congress soon passed legislation expanding the duties of the Service and giving its personnel more authority to address problems. In 1942, all duties and responsibilities relating to vessel inspection and certification of shipboard personnel were transferred by executive order to the U.S. Coast Guard.[42]

On April 16, 1947, a major disaster struck the waterfront of Texas City, Texas. The S.S. *Grandcamp* was taking on a shipment of ammonium nitrate fertilizer at the pier of the Monsanto Chemical Company. A small fire was discovered aboard, but before it could be brought under control, the ship blew up. The explosion instantly killed almost the entire fire department of Texas City, which had responded to the first alarm of fire, as well as all but seven of the ship's crew. Fire and other explosions that followed during the ensuing hours in the waterfront industrial area resulted in the deaths of 468 people, over 2,000 injuries, and property loss of over $67 million. This disaster emphasized the need for regulations in the control of fertilizer-grade ammonium nitrate.

The ports of the nation and the U.S. Coast Guard began vigorous procedures to eliminate the possibility of having vessels containing such materials within their limits.

INDUSTRIAL FACILITIES

Another fire that made an impact on structural fire safety regulations was the fire that occurred in the Triangle Shirtwaist Factory in New York City on March 25, 1911. More than 600 women, most of them young, were working on the eighth, ninth, and tenth floors of this loft building. To prevent unauthorized removal of products, the factory management made a practice of checking purses and bags of all employees as they left the premises. During this procedure the exit doors to the stairs were locked.

The fire started from an unknown cause on the eighth floor at approximately 4:45 P.M. The interior standpipe hoses were rotten and completely ineffective. The fire spread out the windows and onto the floors above. Sixty people jumped to the ground from the eighth, ninth, and tenth floors. A total of 145 people were killed in this disaster and 70 people were seriously injured. The owner of the factory and his family, who happened to be in the building, escaped by going up to the floor above. Charges were brought against the owner as a result of the investigation. He was acquitted because he apparently proved that he was unaware of the practice of locking the stairway exit doors during employee searches.

This fire focused attention on the need for fire safety measures and the safeguarding of occupants in similar buildings. The nation was shaken by this incident, and regulations were established to preclude the possibility of a recurrence. The fire led to enactment of New York State's Labor Law and establishment of the Fire Department's Bureau of Fire Prevention.[43]

The General Motors Transmission Plant fire in Livonia, Michigan, on August 12, 1953, had a major impact on industrial fire safety. This fire, in a building with an undivided floor area of 34.5 acres with only partial sprinkler protection, resulted in the greatest damage from an industrial fire as of that date.

Fire struck the Imperial Foods Processing Plant in Hamlet, North Carolina, on September 3, 1991. The fire's intensity, coupled with several inoperable exits, resulted in 25 fatalities and 54 injuries in this unsprinklered one-story, windowless structure. This fire emphasized the need for adequate means of egress in all industrial plants, including food processors.

HOTELS

Nationwide interest in hotel fire safety was awakened by the fire that swept the Winecoff Hotel in Atlanta, Georgia, on December 7, 1946. This building was widely advertised as a "fireproof" structure and was, in fact, of fire-resistive construction. An open stairway, however, permitted the rapid spread of smoke and heat up the stairs from floor to floor. There were 304 guests in this 15-story building the night of the fire; 119 people were killed, 168 injured. A number of occupants jumped from windows.

A definition of the term *fireproof* from a 1923 publication follows:

A so-called fireproof building bears about the same relation to its contents that a furnace or other stove does to the material put into it to burn. As a rule the fireproof building will prevent the spread of fire to other buildings just as a fire will not spread from one stove to another placed near it; but the contents of a fireproof building will be consumed once the fire is well under way just as thoroughly as the coal and wood in the stove. Further, the heat will be retained in the fireproof building and human beings, if they fail to get out quickly, will be killed.[44]

Reports on the Winecoff Hotel fire indicate that some people who stayed in their rooms were not injured by the fire. They were protected by room doors and managed to obtain air for breathing by opening the windows. Great strides in hotel fire safety came about as a result of this fire. Fire safety improvements were made within the entire state of Georgia, which adopted a fire code soon after the fire.[45]

During the same year, a fire at the La Salle Hotel in Chicago took 61 lives, and 19 lives were lost in a fire in the Canfield Hotel, Dubuque, Iowa. Many similarities existed in the fires in the three hotels.

A 1986 mid-afternoon New Year's Eve fire struck the DuPont Plaza Hotel and Casino in San Juan, Puerto Rico. The fire resulted in 96 fatalities and over 140 injuries. This 20-story, nonsprinklered hotel contained a first-floor ballroom, a second-floor casino, and various mercantile shops, restaurants, and conference rooms. Fire consumed the contents of first-floor areas and was primarily confined to that level. Investigation revealed that the fire had been deliberately set in a large stack of recently delivered furniture that had been temporarily stored in the ballroom.

The government of Puerto Rico convened a commission to study the need for improvements in fire safety code. The commission report brought about improved fire safety standards.

A fire in the MGM Grand Hotel in Las Vegas, Nevada, on November 21, 1980, took 85 lives and injured 600. This fire began in the casino area; however, most of the victims were in the adjacent high-rise section. This fire had an impact on Nevada Code requirements but limited impact elsewhere.

NURSING HOMES, HOSPITALS, AND HOUSING FOR OLDER ADULTS

A number of serious fires have occurred in nursing homes throughout the country. One that focused attention on the field of fire protection and prevention was the Katie Jane Nursing Home fire in Warrenton, Missouri, in 1957. The fire occurred on a Sunday afternoon, when many visitors were in the home, and took 72 lives. Construction deficiencies, the lack of an automatic sprinkler system, and a number of other factors were paramount in this disaster. However, this fire and one that took 63 lives in a Fitchville, Ohio, home for the aged in 1963 did result in the improvement of nursing-home fire safety regulations and procedures in a number of jurisdictions. Until corrective action was initiated at the federal level early in the 1970s, a study of fires indicated a continuing pattern of disastrous fires in nursing homes and related institutions.

Fire safety problems in hospitals were brought to national attention by the St. Anthony's Hospital fire in Effingham, Illinois, on the night of April 4, 1949, in which 74 lives were lost. Combustible interior finish, the lack of automatic sprinkler protection, and general construction were factors in this fire.

A fire that occurred in 1961 in the Hartford Hospital in Hartford, Connecticut, was responsible for 16 deaths. Combustible ceiling tile contributed to the fire's spread, as did the fact that trash and linen chutes in the building opened directly onto the corridors. The structure was a modern fire-resistive building, which proved that such structures are not immune to fires.

In just 13 days in December 1989, fires killed 23 residents in three different housing-for-the-elderly facilities. None of these were licensed or operated as nursing homes.

The first occurred in a six-story, fire-resistive, unsprinklered apartment house for older adults in Roanoke County, Virginia. Four residents died on the third floor, although the fire was confined to the room of origin. The investigation revealed that some of the occupants were difficult to awaken and reluctant to leave their apartments.

Watertown, New York, was the scene of the second fire. Three fatalities occurred, all on the upper floors of this nonsprinklered, fire-resistive apartment building constructed specifically for housing older adults. Smoke movement from the first-floor point of origin was responsible for the fatalities in this December 15 late-night fire.

On December 24, a late-afternoon fire started on the first floor of an 11-story fire-resistive apartment house in Johnson City, Tennessee. This nonsprinklered structure had served as a hotel for many years before conversion to housing for older adults. Sixteen fatalities and 40 injuries occurred. As in the other fires, combustible interior finish was a factor in this incident. Most of the fatalities were on upper floors.

Generally, fire safety requirements in apartments for older adults are no more stringent than those for any other apartment house. Special licensure is normally not required.[46]

SCHOOLS

A six-year-old, three-story school in Colinwood, Ohio, was the scene of tragedy in 1908. The Lakeview School was occupied by over 300 students at the time of a fire. Of those students, 175 were killed. Open stairways and construction materials contributed to the fire spread.

An early school fire brought about a realization that fire safety should be considered even in the smallest of schools. This fire occurred in a two-room elementary school at Babbs Switch, Oklahoma, on December 24, 1924. The school was being used for a Christmas event attended by over 200 community residents. A Christmas tree had been placed in the opening between the two rooms. The windows in the school had previously been barred, and there was only one exit. As a result of these unsafe conditions, 36 deaths ensued in this fire, which was started by a toppled candle on the tree.

FIGURE 1.2 ◆ Station 6, just east of the fire area, served as a hospital for injured firefighters in the 1904 Baltimore fire.

In March 1937, a major explosion struck the New London, Texas, school. This disaster, which killed 297 people, was caused by the improper use of gas. Again, school fire safety was immediately considered to be of paramount concern in communities throughout the country.

School fire prevention and fire protection efforts were greatly enhanced by the tragic fire at Our Lady of Angels School in Chicago on December 1, 1958. This fire, in which 95 lives were lost, was reported to be of incendiary origin. The old multistory structure was of ordinary construction. A delay in turning in an alarm within the building and in contacting the fire department compounded the tragedy. Tremendous efforts to upgrade school fire prevention regulations throughout the country came as a result of this fire. Many schools were provided with automatic sprinkler protection. In addition, a number of states started vigorous fire inspection programs and took other measures to reduce the possibility of fires taking lives because of lack of safeguards.

PRISONS

The Ohio State Penitentiary fire of April 21, 1930, in Columbus, aroused interest in problems relating to fire prevention in prisons. The penitentiary had 4,300 inmates at that time, and 320 were killed in the fire. Most of those killed were trapped in their cells. This fire brought about immediate demands for improvements in fire safety in penal institutions.

Three major prison fires killed a total of 68 people during 1977. Included were 5 fatalities in the federal prison in Danbury, Connecticut; 42 in the county jail in Maury County, Tennessee; and 21 in a St. John, New Brunswick, penal institution. These fires led to new demands for tighter regulation of fire safety within places of incarceration. Interior finish had become a major contributory factor to fires in such places.

CONFLAGRATIONS

Until the 20th century, communities had been built to burn while the country expanded rapidly across the continent. Building conditions in the United States had long been marked by excessive use of combustible materials put together without much regard for safety to life or property from fire. Large individual buildings housing vast amounts of combustible stocks under one roof, lack of fire walls and vertical cutoffs, wood-shingle roofs, and other unsafe factors that contribute to rapid spread of fires were characteristic of the American scene. In many parts of the United States, seasonal droughts and high winds aggravated fire conditions and resulted in area conflagrations that contributed substantially to the high national fire losses. The term *conflagration* usually refers to a major destructive fire.

October 9, 1871, was the day of the two greatest fires in the history of the Midwest. Both started at about the same hour and were compounded by tinder-dry wood resulting from a prolonged drought, plus human carelessness. The better known of the two struck Chicago, taking 300 lives, while structures in an area four miles long and two-thirds of a mile wide were consumed during a two-day period.

The lesser known fire ravaged 2,400 square miles and took 1,152 lives in a forested area of northeastern Wisconsin. This fire is recorded as the Great Pestigo fire because the majority of fatalities occurred in that Wisconsin town.

The "Great Fire of 1901," which burned 146 city blocks in Jacksonville, Florida, on May 3, 1901, was typical of the conflagrations that struck U.S. and Canadian cities during that era. Only seven lives were lost in this fire, which destroyed 2,368 buildings. The fire began in a fiber factory and spread rapidly because of wood-shingle roofs.[47]

On June 6, 1889, the Great Seattle fire swept through what was then downtown Seattle, wiping out 66 square blocks and paving the way for "The Forgotten City" lying beneath the city's modern streets. Because much of the downtown area was on stilts, the sidewalks and streets were great fire carriers. In one instance, firefighters were able to stem the fire north of University Street simply by tearing up the streets and sidewalk and tossing them over the cliff into the bay. To fight the fire, a 200-man bucket brigade was formed along the river that ran past the Olympic Hotel. Water was hauled up from the river and sloshed against the buildings.[48]

Major conflagrations occurring in U.S. cities have also brought reforms in building and fire prevention codes. In 1904, Baltimore was swept by a conflagration that destroyed 80 blocks in the downtown business center. Incoming firefighters from other cities, including New York, were less effective because of incompatible hose threads. This fire resulted in the improvement of construction standards and the development of new procedures in fire prevention. Paterson, New Jersey, lost nearly 500 buildings in a 1902 fire. Other North American cities and towns have had somewhat similar occurrences. For example, an 1861 fire destroyed most of the town of Lindsay, Ontario. Four hotels, two mills, the post office, customs office, and 83 other buildings were destroyed in the town of 2,000. A history of Lindsay states, "The fire spurred the construction of many fine brick buildings to replace the wooden structures which had been consumed."[49]

On October 12, 1918, forest fires swept across northeastern Minnesota, devastating 38 communities and 1,500 square miles. More than 450 people were killed outright, 52,000 more were injured or displaced, and property loss was estimated at $73 million.[50]

In 1923, Berkeley, California, was ravaged by a fire that destroyed 640 structures. In 1961 a conflagration in Los Angeles spread through wooded lands and resulted in the destruction of over 500 dwellings. Some large communities permit the use of wood-shingle roofing. The Los Angeles fire emphasized the need to return to more restrictive code provisions regarding wood-shingle roofing.

Oakland, California, was the site of a devastating fire that took 26 lives and resulted in the loss of 3,469 housing units on October 20, 1991. Property losses were estimated at $1.5 billion. Dry conditions and steep terrain contributed to this loss.

EARTHQUAKES AND FIRE

Although not directly controllable under normal fire safety procedures, the 1906 earthquake and fire in San Francisco must be mentioned in any recount of fires of major significance. The severity of the fires was much greater because of the earthquake. Water mains were broken, which greatly reduced the amount of water available for fire fighting purposes. In addition, the earthquake made response by fire fighting equipment most difficult. A number of fires were uncontrolled.

In this disaster, 422 lives were lost, 28,000 buildings were destroyed, and property loss was estimated at $350 million; almost $7 billion today. This devastating event prompted reconsideration of fire prevention procedures in all areas of the country in which earthquakes might be anticipated. The Baltimore and San Francisco conflagrations led to the development of advanced methods of analyzing risks by the insurance industry.

◆ LABORATORIES IMPROVE FIRE SAFETY

Some of the most important fire prevention efforts start in the laboratory. An example is the work of Dr. Robert C. Kedzie of Michigan. In the early 1870s, residents depended on kerosene oil lamps for illumination. The oil often contained volatile impurities, which caused explosions followed by burnings and death when lamps were ignited.

Kerosene imported into Michigan bore a label "Warranted 150° F Fire Test." This was based on an open-cup test, which did not confine the kerosene vapor to the cup, thereby tending to result in false readings. Dr. Kedzie devised a closed-cup test, which confined the vapors and provided a much more accurate test because it simulated conditions in a kerosene lamp. In many cases the actual flash point was found to be below 100° F with this test. Mandatory use of the closed-cup test has resulted in the saving of countless lives.[51]

Moreover, the foundation of the modern plastics industry—highly flammable products generally yielding toxic fumes—had been laid in the 19th century. The first such substance was Parkesine, made in England in 1866. It was composed of nitrocellulose, camphor, and alcohol. A similar material, celluloid, was manufactured in the United States in 1869. It quickly came into use in everyday items such as shirt collars. A form of this product was used for early motion picture films. Great clouds of yellow smoke billowed out over Ohio's Cleveland Clinic in 1929, when burning nitrocellulose film plates spread noxious fumes throughout the clinic, leaving 125 dead. Most were killed by the fumes rather than by the fire that followed. Safety film was developed soon thereafter.[52]

Another example of improvements attempted in a laboratory setting is the effort to obtain a fire-safe cigarette. The first American patent for a "self-extinguishing" cigarette was issued in 1854. Research to prove the practicability of such a cigarette began in the then National Bureau of Standards in 1932 and has continued through the years as the result of legislative action.

◆ VARIABLES IN THE PHILOSOPHY OF FIRE PREVENTION

The philosophy of fire prevention includes many variables. The term *fire prevention* varies in interpretation within the fire protection field. There had been a tendency to include some closely related activities under fire prevention responsibilities. Some of these activities are not truly preventive in nature.

As an example of added meanings, a number of activities that are actually *fire reactions* are considered to be "fire prevention practices" by both lay and professional people. The practice of home and office fire drills, for example, is usually associated with fire prevention programs even though it is actually a fire reaction type program. Probably one reason for this is that the same individuals generally promote both programs. A definition of prevention that seems to fit: "Prevention is an assertive process of creating conditions and/or personal attributes that promote the well-being of people."[53]

Certain fire prevention concepts are not strictly related to the prevention of fire but are more closely related to the *prevention of the spread of fire*. For example, wearing noncombustible clothing will not prevent the ignition of a match, but wearing such

FIGURE 1.3 ◆ Passing years too often erase the memory of firefighters who gave their lives at fire scenes.

clothing does retard the possible spread of a fire that might be started on the clothing by the match. The personal steps taken to reduce fire spread possibilities on clothing are one aspect of fire prevention. However, measures taken to preclude the possibility of starting a fire are more truly the essence of fire prevention.

Another fire prevention measure is making sure that fires that might occur in a structure will not entrap individuals within the building. People who happen to be in the structure at the time of the fire should have every opportunity to leave the building safely. An important factor here is individual conditioning to fire reaction, in which health, physical abilities, past exposure to fire, and many other variables play a part. The term *fire safety* is seen by many as encompassing fire prevention, fire reaction, and prevention of fire spread.

◆ **SUMMARY**

History teaches us much about fire prevention. It has shown us that virtually all programs used in the area of fire prevention are the result of a disastrous fire. This aspect has not changed through the years. In spite of improved technology, science, and the efforts of many fire prevention professionals, it almost always requires a large loss of life or property to affect significant change in fire prevention practices. In addition, a review is necessary of some of the significant contributing factors, whether they be blocked exits, inward opening doors, flammable or combustible finishes, lack of alarms, or lack of automatic sprinklers. We can also note that politics and business (profits and the bottom line) greatly affect fire prevention. Having the ability to find a solution to fire risks is not enough. It takes leadership and a willingness to influence those with the power to make the necessary changes. The danger of fire to life and property has always been present since its discovery. Due diligence of fire prevention professionals must be constant.

■■■

Review Questions

1. A curfew originally meant
   a. setting time to vacate public areas
   b. a time to extinguish fire
   c. getting children off the streets
   d. establishing corporal punishment for fire prevention violations
2. Early fire prevention efforts included
   a. construction requirements
   b. behavior standards and cases
   c. maintenance of fire safety devices
   d. punishment for violators
   e. all of the above
3. Fire inspections in the New World probably began in
   a. New Amsterdam
   b. Boston
   c. Williamsburg
   d. Plymouth Rock
   e. Philadelphia
4. Which was not a topic of discussion in the first annual conference of the National Association of Fire Engineers in 1873?
   a. repression of incendiarism
   b. fire escapes
   c. automatic sprinklers
   d. reduction of excessive height of buildings
   e. fire investigations for cause
5. In 1913, Milwaukee had how many employees devoted to fire prevention?
   a. 5
   b. 10
   c. 15
   d. 20
   e. 30
6. Common contributing factors to the disastrous fires at the Iroquois Theater, Rhythm Club, and Coconut Grove include
   a. inadequate exits
   b. inward opening doors
   c. flammable decorations
   d. all of the above
7. Which fire contributed to the beginning of labor laws for the safety of workers?
   a. General Slocum
   b. Triangle Shirtwaist
   c. General Motors Transmission Plant
   d. Beverly Hills Supper Club
   e. MGM Grand Hotel
8. A school fire in which city contributed the most to today's fire prevention efforts in schools?
   a. Colinwood, Ohio
   b. Babb's Switch, Oklahoma
   c. New London, Texas
   d. Chicago, Illinois
   e. New York, New York
9. Which two great conflagrations started the same day?
   a. Chicago and Pestigo, Wisconsin
   b. Chicago and Seattle
   c. Seattle and Baltimore
   d. Baltimore and San Francisco
   e. Berkeley and Oakland
10. Fire prevention or fire safety includes
    a. fire reaction
    b. fire prevention
    c. prevention of the spread of fire
    d. all of the above

■■■

Answers

1. b
2. e
3. a
4. c
5. e
6. d
7. b
8. d
9. a
10. d

Notes

1. Historical references to early Roman and English periods drawn from *A History of the British Fire Service* by G.V. Blackstone, CBE, G.M. (London: Routledge & Kegan Paul, 1957).
2. Ibid., p. 10.
3. Alexander Reid, *"Aye Ready!"* (Edinburgh: Geo. Stewart & Co., 1974), p. 5.
4. Blackstone, p. 17.
5. Ibid., p. 33.
6. Ibid., p. 87.
7. *Project 9: The Story of Fire Fighting* (London: The Home Office and the Central Office of Information, 1976), p. 1.
8. Blackstone, p. 84.
9. Ibid., p. 87.
10. Harry Klopper, *The Fight Against Fire* (Birmingham, England; Birmingham Fire and Ambulance Service, 1955), p. 36.
11. Paul R. Lyons, *Fire in America!* (Boston: National Fire Protection Association, 1976), p. 2.
12. Charles L. Radzinsky, *100 Years of Service 1872–1972* (Rensselaer, N.Y.: Hamilton Printing Company, 1972), p. 9.
13. Edward R. Tufts, *A History of the Salem Fire Department* (Salem, Mass.: Holyoke Mutual Insurance Company in Salem, 1975), p. 3.
14. *History of the Fire Department, Norfolk, Virginia* (Norfolk, Va.: Norfolk Fire-fighters' Association, 1975), p. 32.
15. Ibid., p. 8.
16. James C. Mullikin, *A History of the Easton Volunteer Fire Department* (Easton, Md.: Easton Volunteer Fire Department, 1962), p. 9.
17. The National Commission on Fire Prevention and Control, *America Burning* (Washington: The National Commission on Fire Prevention and Control, 1973), p. 79.
18. Patrick T. Conley and Paul R. Campbell, *Firefighters and Fires in Providence* (Providence, R.I.: Rhode Island Publications Society, 1985), p. 3.
19. *Reading's Volunteer Fire Department,* comp. Federal Writers Project of Works Progress Administration in the Commonwealth of Pennsylvania, Berks County Unit (Philadelphia: William Penn Association, 1938), p. 6.
20. Ibid., p. 7.
21. Ibid.
22. Arnold Rosenbleeth, *Firefighting in Pensacola* (Pensacola, Fla.: Pensacola Historical Society Quarterly, Winter 1980), p. 2.
23. Leo E. Duliba, *A Transition in Red* (Merrick, N.Y.: Richwood Publishing Co., 1976), pp. 12–13.
24. *Greensboro Fire Department, 1808–1984* (Dallas, Tex.: Taylor Publishing, 1984), p. 6.
25. "Past and Present History Denver Fire Dept.," unpublished paper, Denver, Colo., p. 2.
26. Lyons, pp. 29–30.
27. Perry H. Merrill, *Montpelier, the Capital City's History* (Montpelier, Vt.: published by author, 1976), p. 62.
28. Arthur A. Hart, *Fighting Fire on the Frontier* (Boise, Idaho: Boise Fire Department Association, 1976), p. 33.
29. Betty Richardson and Dennis Henson, *Serving Together: 150 Years of Firefighting in Madison County, Illinois* (Collinsville, Ill.: Madison County Firemen's Association, 1984), p. 45.
30. *The Fire History of the City of West Palm Beach* (West Palm Beach, Fla.: West Palm Beach Fire Department, 1980), p. 6.
31. Frank and Gennie Myers, *Memphis Fire Department* (Marceline, Mo.: III Walsworth, 1975), p. 30.
32. *Tulsa Fire Department 1905–1973* (Tulsa, Okla.: Intercollegiate Press, 1973), p. 17.
33. Donald M. O'Brien, *The Centennial History of the International Association of Fire Chiefs* (n.p., 1973), p. 6.
34. Ibid., p. 43.
35. R.L. Nailen and James S. Haight, *Beertown Blazes: A Century of*

Milwaukee Fire Fighting (Milwaukee, Wisc.: NAPCO Graphic Arts, 1971), pp. 21–22.

36. *Long Beach Fire Department* (Long Beach, Calif.: Long Beach Fire Department, 1976), p. 29.

37. *Phoenix Fire Fighters* (Phoenix, Ariz.: Phoenix Fire Department, 1983), p. 23.

38. Houston Fire Museum, Inc., *Houston Fire Department 1838–1988* (Dallas, Tex.: Taylor Publishing, 1988), p. 13.

39. Sharon K. Brace, *Fire on the River* (Inglefield, In.: APS Publishing, 1995), pp. 28–29.

40. Bob Considine, *Man Against Fire* (Garden City, N.Y.: Doubleday and Co., 1955), p. 134.

41. *Great Fires of America* (Waukesha, Wisc.: Country Beautiful Corporation, 1973), p. 134.

42. Walter C. Capron, *U.S. Coast Guard* (New York: Franklin Watts, Inc., 1965), pp. 43–44.

43. *WNYF* (New York: New York Fire Department, 3rd issue, 1980), p. 20.

44. C. C. Dominge and W. O. Lincoln, *Fire Insurance Inspection and Underwriting* (New York: The Spectator Co., 1923), p. 323.

45. Sam Heys and Allen B. Goodwin, *The Winecoff Fire* (Marietta, Ga.: Longstreet Press, 1993).

46. *Alert* (Bulletin No. 90–1, Quincy, Mass.: National Fire Protection Association, 1990).

47. James Robertson Ward, *Old Hickory's Town* (Jacksonville, Fla.: Old Hickory's Town, Inc., 1985), p. 175.

48. "The Great Seattle Fire," *Wildlife News Notes,* NFPA, June 1998.

49. Alan R. Capon, *Historic Lindsay* (Belleville, Ont.: Mika Publishing, 1974), p. 18.

50. Francis M. Carroll and Franklin R. Raiter, *The Fires of Autumn* (St. Paul, Minn.: Minnesota Historical Society Press, 1990).

51. George P. Merk, "George C. Kedzie, Michigan's Nineteenth Century Consumer Advocate," *Michigan History* (Lansing, Mich.: Bureau of History, Secretary of State, January/February 1989), pp. 20–21.

52. Richardson and Henson, p. 56.

53. William A. Lofquist, *Discovering the Meaning of Prevention* (Tucson, Ariz.: AYD Publications, 1983), p. 2.

Status of Education, Engineering, and Enforcement in USA

In this chapter the challenges in fire prevention as they exist in the early part of the 21st century will be discussed. In Chapter 1 problems and solutions found in earlier centuries were noted. The majority of these solutions still have application today. The proceedings of the *Official Record of the First American National Fire Prevention Conference* published in 1914 contains many fire prevention concepts that are just as applicable today as they were in the early 20th century.[1] Likewise, subsequent national gatherings related to the subject contain similar applicable suggestions.

◆ THE THREE "Es"

Even today the basic concepts of Education, Engineering, and Enforcement apply to the subject of fire prevention. An equation might be Education + Engineering + Enforcement = Fire Safety. At this point it might be in order to add yet another definition of fire prevention to those offered in Chapter 1. The *Municipal Fire Service Workbook* definition follows: "Prevention can be defined as the effort to decrease the chances of unwanted ignition and, to some extent, to limit the spread of fire by methods which are independent of actions taken after ignition occurs." This definition is designed to include fire prevention activities carried out by municipal fire departments but not fire prevention which involves the actions taken by the fire department to bring the unwanted fire under control.[2]

The authors of the preceding definition point out in a footnote that "The line of demarcation between prevention and suppression is therefore somewhat fuzzy. A case in point is the installation of fire walls. Clearly the walls have little to do with chances of fire ignition. Their main function is to limit the spread of fire for a specified period of time to limit fire loss until the suppression force arrives."

Much progress has been made in the field of fire safety education. A much higher percentage of children are exposed to this subject than in earlier years. In many cases fire safety education is folded into overall safety classes offered at local schools.

FIGURE 2.1 ◆ Firefighters are very effective teachers in classrooms. *(Grand Lake, Minnesota, Fire Department)*

Unfortunately, the offering of such classes is often left to the discretion of the local fire department, and some do not see fit to offer this service to the community.

Education, engineering, and enforcement will be addressed in this chapter. Before delving into those phases of fire prevention it is best to take a look at what we mean by all three of these concepts. Let us first take a look at the problems we are facing today. Greater emphasis on fire prevention as opposed to suppression as a fire department function is still needed but much progress has been made. It is rare to find a community with a population of more than 10,000 that does not have at least one person in the fire department, whether career or volunteer, assigned to fire prevention. Usually that one person in a smaller community is attempting to address all of the Es with the remainder of the force devoting its time primarily to suppression duties.

A study commissioned by the U.S. Fire Administration in 2002 found that only four cities with a population of over one million had no selected (recognized) fire prevention or code enforcement programs in effect. As the population of cities dropped below one million, the percentage of departments lacking such programs increased substantially. The study concluded that 62 percent of fire departments in small cities and towns had no established programs. Sadly, these departments protect some 29 percent of the population of the United States.[3]

Areas of responsibility for fire prevention–related duties have become more specialized through the years. For example, more fire departments have staff who are trained specifically for public fire safety education assignments. This is especially true in larger fire departments. Plan review and fire investigation, factors often overlooked as fire prevention measures, are more often specialized assignments. Certainly both contribute materially to the prevention of fire. Abatement of arson through thorough investigation is a major contributor to fire prevention. Likewise, plan review by a trained fire protection specialist can result in a decrease in the loss of life and property in a community.

All three of the three Es have seen a higher degree of specialization in the way they are being administered. The idea of having college-trained fire protection engineers involved in reviewing construction plans and specifications for governments and the private sector was only a dream when the first college-level fire protection engineering program opened in the early 1900s. Mass media concepts such as television and radio were not available to disseminate fire safety education concepts at that time. Probably the strongest application of one of the Es was in the enforcement realm. Many examples of this phenomenon are found in Chapter 1.

◆ EDUCATION

As previously noted, fire safety education has become much more widespread in recent years. Smoke alarms are now found in most homes. An increasingly higher percentage of smoke alarms are being electrically connected as a result of requirements for new dwellings, thereby reducing the need to rely on the consumer changing batteries each year. Smoke alarms are estimated to be responsible for cutting home fire fatalities in half.

Although residential sprinklers are slowly gaining acceptance, they have yet to make a major impact on home fire fatalities because they are installed in so few homes.

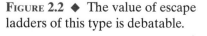

FIGURE 2.2 ◆ The value of escape ladders of this type is debatable.

There is no doubt they will eventually have a profound impact on fire safety in dwellings. This concept is discussed in greater detail in Chapter 4. Sprinklers have proven to be real lifesavers in a number of occupancies, and private homes will someday be added to that list.

"Stop, Drop, and Roll" has become a well-practiced option to running with burning clothes. The advent of child-resistant lighters has also had a positive effect on home fire safety. However, smoking coupled with alcohol continues to be a major contributor to home fire fatalities. Although New York State has enacted legislation mandating the self-extinguishing cigarette, efforts at the national level in the United States continue to meet fierce opposition. Measures to reduce the combustibility of mattresses and other bedding material have been helpful, especially in institutional occupancies.

The 20th century has seen the introduction of hundreds of new chemicals and related hazardous materials, many of which create code enforcement and suppression problems for the fire service. Specialized hazardous materials response units abound in the fire service. The everyday use of many hazardous materials has become a necessity of modern life and fire codes have had to adjust to these changes. For example, Danforth's fluid, which was once banned in major cities, is now in common use. The popular name of the product is "gasoline." The advent of self-service gasoline stations during the latter part of the 20th century is another example of a procedure long banned but now recognized in codes because of consumer and operator demand for this method of operation. Liquefied natural gas is another product once prohibited but now commonly used.

The reintroduction of common household appliances such as the wood-burning stove have likewise contributed to the fire problem in some regions. Again, fire and building codes have had to be adjusted to recognize the popular demand for such products. Clearly, factors other than safety figure in the establishment of code criteria. Code developers have been forced to "give in" to certain consumer-related issues. Often these changes are opposed by fire and building officials but they are overpowered by other factors.

Although fire safety in nursing homes has been greatly improved, 15 were killed in a Hartford, Connecticut, facility in February 2003 followed by 15 fatalities in a similar facility in Nashville, Tennessee, in September 2003 and 5 in a Maryville, Tennessee, nursing home in January 2004.[4]

Greater examination of specific needs in the fire safety education field is another trend that has developed in the latter part of the 20th century and is still prevalent in the early part of the 21st century. Individual groups are now being spotlighted for special attention geared to their needs. This trend is replacing the broad brush approach to public education used in the past.

Such a procedure is exemplified by the 2002 U.S. Fire Administration report entitled *Babies and Toddlers Fire Death Data Report*.[5] This report concludes that children under 5 years of age are twice as likely to die from fire than the rest of the population. The document includes data from all state fire marshals indicating the trends in their individual states. The publication shows the ranking of fire deaths for children under 5 compared to other causes of death for this population group in each state. It also includes comparisons for these fatalities with those of several other countries. Public fire safety educators and others interested in the field are strongly encouraged to develop and carry out programs which will create a greater awareness among parents and guardians of actions they can take to preclude such horrific events.

There is no doubt that improved application of building and life safety codes, increased use of fire safety education, installation of smoke alarms in residences, and other measures initiated by the fire service and allied organizations have done much to reduce fire occurrences and most particularly fire fatalities during the last 10 to 15 years of the 20th century. The advent of the U.S. Fire Administration and its National Fire Academy has also done much to improve fire safety in the United States.

A review of fire fatality statistics of the early years of the 20th century provides a striking comparison with those of the present era. In 1913 the U.S. fire fatality death rate was 9.1 persons per 100,000 population, while in 2002, just 90 years later, it had dipped to 1.4 per 100,000 population. Canadian fatality rates are slightly higher in both year groupings. While this represents a drop of nearly 400 percent in less than 100 years, it must be recognized that the accuracy of compiling fire fatality statistics is greater today than in 1913. It is not clear how this might affect the death rate statistics.[6]

A U.S. Fire Administration report issued in 1998 indicates that fire death rates are 36 percent higher in rural than nonrural areas. Fire death rates for African Americans living in rural areas were over three and one-half times as high as for rural whites. Fire deaths for men, regardless of location of residence, are twice as high as for women. Heating devices, careless smoking, and electrical distribution are the top three causes for fires in rural areas. The lack of working smoke alarms in manufactured homes is a special problem in the South.[7]

In a 2004 report by the American Burn Association, 4,500 fire and burn deaths per year are estimated. This includes an estimated 750 from automobile and aircraft crashes. The total also includes contact with electricity and chemical products, sources often not identified as fire fatalities. Nevertheless, the total represents a trend downward in this category. Likewise a drop from 2 million annual burn injuries treated to over 1 million during a recent 10-year period represents a major decline.[8] Many burn deaths are not reported as "fire" deaths.

◆ ENGINEERING

The second and third of the Es, Engineering and Enforcement, are addressed in the codes in use to assure an acceptable degree of fire safety. It is presumed that engineering principles are employed to bring about this condition. Closer examination will reveal that in some cases the Delphi principle is used to develop requirements in codes. This principle entails the reaching of the "best solution" through discussion by a group of experts in the subject matter.

While major attention has been focused on the World Trade Center catastrophe, the Pentagon incident in Arlington, Virginia, has more serious implications for the average city. Once an airport is established as a major hub, development flourishes around it. The possibility of an errant aircraft (not necessarily flown by terrorists) striking a large building adjacent to an airport exists in many cities in North America. Major airports serving Baltimore, Chicago, Las Vegas, Los Angeles, New Orleans, Toronto, and Washington present this potentially dangerous situation.

A late 20th century and early 21st century concept (as exemplified by the tragedies of September 11, 2001) is the question of whether construction of high-rise structures should include an ability to withstand the impact of a large airplane filled to capacity with flammable fuel. While this question has a direct bearing on the construction of the building from a fire safety standpoint, it travels far beyond the perils of a normal fire occurrence in such a structure, even a fire started by an arsonist working inside the structure. This issue is being debated today and has more application than solely the potential for terrorists to employ this method of destruction.

Failure to include details of either the New York or Arlington, Virginia, event of September 11, 2001, is in no way intended to downplay the extreme importance of either, but the vast amount of media and fire service coverage given to both precludes the need for duplicative coverage in this text. These events emphasize the major role the fire service has in attempting rescue and fire control in terrorist incidents. Many of the same fire service personnel who inspect these type of properties are involved in evacuation attempts and fire suppression.

An October 17, 2003, fire in the 35-story Cook County Administration Building in Chicago took six lives in this nonsprinklered building. As many as 2,500 persons may have been in this structure during business hours.[9] Office occupancies have long been considered among the safest in that occupants are alert and able to respond to fire drill procedures. This fire exemplifies the fact that this type of occupancy still presents a significant challenge.

It is not unusual for six persons to die in a home fire, an event that makes headlines only in the local media, while the office building fire garners national attention. Major progress is being made in reduction of home fire fatalities. The National Fire Protection Association (NFPA) reports the death toll in home fires at 57 percent less in 2002 than in 1977. In fact 2,670 fatalities in home fires were reported by NFPA in 2002, the lowest number ever recorded by that organization. In most jurisdictions around three-quarters of fire responses are to dwelling fires. Most of the remaining fire fatalities occur in motor vehicles.[10]

The vast majority of all fires are in places where most people feel safe—in their homes and automobiles. Of course, the fire service has little direct control over motor vehicle fatalities except through child seat safety programs and emergency medical services responses. Other occupancies in which people sleep, such as nursing homes, hospitals, hotels, and other residential facilities, have become much safer from a fire safety standpoint through the years.

DEVELOPMENT AND ENACTMENT OF FIRE SAFETY CODES

Fire prevention codes, which have historically been concerned with fire safety issues that relate to fire protection equipment, maintenance of buildings and premises, and of hazardous materials, processes, and machinery used in buildings, are generally administered and enforced by the fire department.

Building codes, which are usually administered and enforced by the building department, address fire safety requirements with respect to the construction of buildings. Because there may be some areas of overlap, a cooperative relationship between the fire department and the building department is essential for the good of the community. It should be noted that neither fire prevention nor building codes address furnishings of dwellings, which are major contributors to fire fatalities.

Generally, fire safety and building codes, when adopted for the first time in a state or community, do not apply to existing structures or installations except when the enforcement official determines that continuation of the hazard would jeopardize safety of life and property. The inclusion of such a clause does not exempt existing facilities from code application; it merely places an onus on the enforcement official to be in a position to justify retroactive application from a life and property safety standpoint.

Fire prevention requirements in some form or another have been a part of our society for many generations. The inclusion of a document known as a fire prevention or fire safety code in the municipal government framework can be traced in more recent years to the influence of the insurance industry on public fire protection. The National Board of Fire Underwriters, in the aftermath of the great Baltimore fire of 1904, developed the Standard Grading Schedule[11] for cities and towns with reference to fire defenses. This had a profound effect on the development and enactment of municipal fire prevention codes.

The National Board of Fire Underwriters (later known as the American Insurance Association) also formulated a model Fire Prevention Code, copies of which were made available to jurisdictions for the asking. Countless localities throughout the United States adopted the Code without change as the legally constituted fire prevention code. Although some cities probably would have developed and implemented fire prevention codes on their own, it is doubtful that there would have been anywhere near the use of fire prevention codes without the availability of this Code.

MODEL FIRE PREVENTION CODES

The National Fire Protection Association (NFPA) Uniform Fire Code covers subjects typical of those covered in a modern state, county, or municipal fire prevention code. This Code, which is adopted in a number of jurisdictions, has chapters covering administration and enforcement; general fire safety requirements; means of egress; fire protection systems and equipment; automatic sprinkler systems; fire detection and alarm systems; occupancy fire safety requirements, ranging from assembly occupancies to airports and heliports; and special processes and material handling ranging from hazardous materials and chemicals to safeguards during building construction and demolition operations. The Code also contains a definitions chapter and an appendix. As with the other model codes, updated editions are developed every few years.

A fire prevention code prepared by the International Code Council was designed to be used in coordination with the International Building Code. Many communities using the Building Code adopted this fire prevention code.

MODEL BUILDING CODES

Building codes play an important part in the protection of the community from a fire safety standpoint. Code coverage includes structural requirements, fire protection as related to structural elements, means of egress, interior finish, vertical and horizontal openings, and many other areas that are directly related to fire protection. The code official must by necessity be concerned with all fire prevention and fire protection requirements with respect to original construction. Unless building officials and fire department personnel enjoy a close working relationship, possible conflicts of authority could be detrimental to community service and safety.

Of all departments in city government, the building department works most closely with the fire department. Like all municipal departments, the fire fighting services need certain tools to perform their duties—fire trucks, hoses, ladders, etc. However, one of the least recognized tools used by the fire department is the city building code or ordinances. Too many municipalities still have inadequate ordinances that give the fire departments little or no authority to require adequate fire protection measures to reduce fire losses.[12]

Figure 2.3 ◆ Water storage tanks such as this one at the Madison County Court House in Florida are important elements of public fire protection.

The International Codes. In 1994, the three nationally recognized model code organizations—BOCA, ICBO, and SBCCI—formed the International Code Council (ICC). The ICC is a nonprofit umbrella organization established with the express purpose of developing a comprehensive and coordinated set of model construction codes called the *International Codes.* The International Plumbing, Private Sewage Disposal, Mechanical, Fuel, Gas, Zoning, and Property Maintenance codes have been produced. These codes and the International Building, Fire Prevention, and Residential Codes have been developed and are available for adoption.

The International Building Code contains a number of chapters which relate to fire protection: use or occupancy, building heights and areas, type of construction, fire resistant materials and construction, interior finish, fire protection, means of egress, accessibility, roof assemblies and rooftop structures, mechanical systems, and a number of other subject areas having some bearing on this subject.

Several major cities in the United States utilize locally developed building and fire codes. Their officials find these codes more suitable for conditions unique to these metropolitan jurisdictions.

National Fire Protection Association Standards. Both the model building and the fire prevention codes adopt by reference selected National Fire Protection Association (NFPA) standards as portions of the code. Among the more popular standards adopted by reference are Flammable and Combustible Liquids Code (NFPA No. 30); Liquefied Petroleum Gas Code (NFPA No. 58); National Electrical Code (NFPA No. 70); and the Life Safety Code (NFPA No. 101). In a few cases there are conflicts between the specific provisions of the individual building code and the NFPA standard adopted. An example is the Life Safety Code (NFPA 101), in which travel distances to exitways and other specific requirements do not always coincide with all building code requirements. The NFPA also publishes a building code known as the *Building Construction and Safety Code* (NFPA No. 5000).

Mini–Maxi Codes. The concept of the mini–maxi codes has caused a great deal of concern in fire service circles. Under this concept, municipalities and counties within a state adopting such a code are prohibited from adopting codes that are less stringent or more stringent than the state code. This approach gains uniformity of application but is seen by many as inhibiting a local jurisdiction from adopting more stringent fire protection requirements, such as stronger sprinkler provisions, which may be needed to address unique conditions.

Two examples of the manner in which mandatory state building codes with no local amendment potential are inimical to effective fire safety are noted. The first involved a fraternity house fire in Chapel Hill, the home of The University of North Carolina. This 1996 fire took five lives in a three- and four-story nonsprinklered house built in 1927. Because the town could not require such protection under the state's mini–maxi code, it was necessary to obtain enactment of state legislation exempting Chapel Hill from these provisions in order to enable the promulgation of a retroactive sprinkler ordinance.

The other example was in a state with a mini–maxi code which prohibited local jurisdictions from requiring the installation of AC-powered smoke alarms in existing apartments. A family that had recently moved to the United States and whose members were not familiar with smoke alarms of any kind were killed in a fire in their apartment, which had a nonworking alarm. Subsequent to the fire, inspectors found many such alarms in the same building.

Performance-Based Codes. *To prescribe* means to "lay down as a guide or rule of action." *Performance* means "the act or process of performing." Building and fire safety codes have long been prescriptive in composition with details of what needs to be done in order to comply. Specific dimensions, for example, are included. The code enforcement official has defined parameters within which to work as has the architect, contractor, and user.

Building Code Coverage
Both the *International Building Code,* published by the International Code Council, and the *Building Construction and Safety Code,* of the National Fire Protection Association, give primary credence to the potential classification of use and occupancy when evaluating construction criteria for a building. The categories addressed in the International Code—the present iteration of the former BOCA, ICBO, and SBCCI codes—are as follows: Assembly, Business, Educational, Factory, High Hazard, Institutional, Mercantile, Residential, Storage and Utility, and Miscellaneous. The original use of the building is a major factor in design criteria. There are breakdowns within the noted classifications; for example, there are five different types of Assembly occupancies, ranging from Motion Picture Theaters to Stadiums.

The NFPA code has a more detailed classification of use and occupancy, with a total of 15 basic categories as opposed to ICC's 10; however, the International Code has special coverage for unusual or special structures such a high-rise buildings and covered malls. The NFPA also has specific coverage for a number of special structures. The use and occupancy classifications used by NFPA closely track the categories utilized in the International Code.

Once the use and occupancy of the structure is determined, the next step for the design professional is to check out the construction-type limitations for the type of use and area of the building under design. In each type of construction there is a height and area limitation for each use and occupancy. For example, a four-story frame nursing home (institutional occupancy) could not be built under either of the two nationally recognized codes.

Under the ICC code are five types of construction. Types I and II construction in which elements of construction are noncombustible; Type III in which exterior walls are noncombustible and the interior is of some other permitted type of construction; Type IV in which exterior walls are noncombustible and interior building elements are heavy timber or other substantial wood materials without concealed spaces; and Type V in which any other material permitted by the code is used.

The NFPA code breaks out construction types basically the same as in the ICC code; however, Type V is specially designated as wood or other approved material. The method of designating use and occupancy as opposed to construction criteria is slightly different in the NFPA document.

The NFPA code has a section for limited coverage of the Performance concept of obtaining code compliance while the ICC publishes a separate code entitled *Performance Code for Buildings and Facilities*. NFPA's coverage is primarily in the Safety to Life realm while the ICC publication addresses a wider range of factors relating to building construction.[13]

Performance-Based Design. By definition, performance-based design is an engineering approach to design elements of a building or facility based on performance goals and objectives, engineering analysis, scientific measurements, and quantitative assessment of alternatives against the design goals and objectives, using accepted engineering tools, methodologies, and performance criteria.

Performance-based design provides a new set of challenges for the Authority Having Jurisdiction (AHJ). Rather than requiring the building to meet a list of prescriptive requirements, the AHJ must evaluate how the structure and its occupants will perform under fire conditions. This means the AHJ must be familiar with principles of fire behavior, structural performance, human response, and integrated life safety and fire protection systems.

A statement in an analysis of performance-based codes developed by the Boston Fire Marshal notes the three sources of information for the plans reviewer utilizing this method: (1) a set of objectives, (2) a design guide, and (3) reference material. The report states that: a code official trying to use these documents to ensure the safe design of buildings is analogous to a police officer trying to enforce a safe society by using books on philosophy and theology. These books may contain valuable information as to how one should conduct affairs but are also useless as a set of enforceable rules.[14]

Building and fire prevention codes in the United States and Canada have been developed entirely through a prescriptive method in past years. This means that the codes state that corridors, for example, must be at least a certain width in order to be in compliance. Widths and other code features were often established by tests conducted by various agencies and organizations.

Many of the research projects that influenced code requirements were conducted by Underwriters Laboratories, Inc. and the National Bureau of Standards, the forerunner of today's National Institute of Standards and Technology. Tests that determined fire resistant rating categories were first conducted by the National Bureau of Standards at a large government building slated for demolition in the nation's capital, Washington, D.C.[15]

The "prescriptive type code" designates specific parameters for construction which makes it a rather simple task to discern deviations. Several other nations—Australia, Canada, New Zealand, Sweden, Japan, and China—have found that the so-called "performance-based code" provides greater latitude for the design professional to fashion a structure embracing the specific needs of the client without having to adhere so closely to predetermined limitations as imposed by a prescriptive type code. [16]

Performance-based design, as defined by a Federal Emergency Management Agency publication, is an engineering approach to design elements of a building or facility based on performance goals and objectives, engineering analysis, scientific measurements, and quantitative assessment of alternatives to meet the design goals and objectives, using accepted engineering tools, methodologies, and performance criteria.[17]

It is a difficult task for the code enforcer to determine whether the design professional (architect/engineer) has developed performance-based criteria that will enable the structure to withstand the ravages of fire. The code enforcer must know a great deal about suppression system capabilities, human reaction capabilities, and anticipated structural integrity retention in the event of fire. In practically all cases a combination of these issues must be considered in making a final decision as to the appropriateness of the design proposal. If any portion of the system fails there is only the "good judgment" of the code enforcer to rely upon as no specific code violations may have contributed to the incident. Such an incident may well leave the fire and/or building code official "hanging out to dry" with no real avenues of defense for his or her decision as there is with a prescriptive code. Generally the code official has the option of accepting use of a performance-based code or not. In some cases a combination of prescriptive and performance codes might be utilized.

Many fire and building code enforcement officials do not feel comfortable administering the application of a performance-based code. They feel that they are on much firmer ground in using a prescriptive type code. The International Code Council has developed a *Performance Code for Buildings and Facilities*. In addition, the National Fire Protection Association has produced NFPA Standard 101A, *Guide to Alternative Approaches to Life Safety*. This standard was originally developed with assistance from the Health Care Financing Administration, a federal agency, with a goal of providing a means of evaluating existing nursing homes and health care facilities through consideration of a number of fire safety measures and programs.

The chart that follows[18] lists some of the major deficiencies found in ten major U.S. fires, all of which had profound effects on fire and building code safety enhancements and enforcement techniques. In some cases the losses represented failures in application or enforcement of codes that existed at the time, while others were brought about by shortfalls in existing codes. Several of these fires resulted in improvements of codes that were in existence at the time of the fire. The Fitchville, Ohio, Nursing Home fire in 1963, for example, played a major role in federal action to mandate improved fire protection features in health care facilities housing patients receiving aid from the Medicare and Medicaid programs. There were 63 fatalities in this fire.

Fire Incident	*Date*	*Lives Lost*	*Predominant Code Issue*
Iroquois Theater Chicago, Illinois	12/30/1903	602	• Exit doors swing in direction of egress • Fire retardant decorations
Triangle Shirtwaist Factory New York, New York	3/25/1911	145	• Inadequate and locked exits • Accumulations of combustibles
Cleveland Clinic Cleveland, Ohio	5/15/1929	125	• Storage of cellulose nitrate films
Ohio State Penitentiary Columbus, Ohio	4/21/1930	320	• Inadequate egress supervision • Lack of fire protection equipment
Coconut Grove Boston, Massachusetts	11/28/1942	492	• Number of exits • Exit door swing • Combustibility of decorations
LaSalle Hotel Chicago, Illinois	6/5/1946	61	• Corridor protection • Enclosure of vertical openings
Winecoff Hotel Atlanta, Georgia	12/7/1946	127	• Enclosure of vertical openings • Early notification (fire alarm) • Recognition that "fireproof" does not exist
GM Transmission Plant Livonia, Michigan	8/12/1953	3	• Concealed combustible spaces • Unprotected steel columns • Lack of fire separations • Lack of sprinklers • Large quantities of heated combustible liquids
Hartford Hospital Hartford, Connecticut	12/8/1961	16	• Combustible ceiling tile and glue • Delayed fire reporting • Open doors onto corridors
Golden Age Nursing Home Fitchville, Ohio	11/23/1963	63	• Lack of fire separations • Lack of fire protection equipment

ZONING CODES

Zoning codes also have an effect on fire prevention in a community. Through proper zoning, various types of occupancies are limited to given sections of the community. Bulk storage of flammable liquids would not, under a zoning code, be found in the middle of the high-value mercantile district. This condition does exist in a number of communities over the country; however, in practically all cases, flammable liquids storage was set up at the location before the advent of zoning or fire prevention regulations.

Zoning provisions also help in making sure that adequate clearances are provided across streets. This affects the potential for a conflagration in the community. Explosives storage and manufacturing are likewise generally very tightly controlled by local zoning requirements.

ELECTRICAL CODES

Electrical codes are primarily fire prevention codes in that electrical safety is so closely related to fire prevention. The principal purpose of electrical inspections is to see that wiring is safely installed in such a manner that people in the structure will not be directly endangered and that fires will not be started as a result of faulty installation.

There is a greater degree of uniformity of code usage in the electrical field than in any other field of public safety code coverage. The National Electrical Code, developed by the National Fire Protection Association, is incorporated as the electrical code in practically every state, as well as in practically all local jurisdictions having an electrical code in effect. Methods of enforcement vary; however, municipal electrical inspection is probably the most prevalent method employed.

In some areas, electrical code enforcement is at the state level, while in a few areas this responsibility is carried out by a private inspection organization. Work arrangements and procedures followed are essentially similar to those of any public electrical inspection agency.

Electrical inspections, like building code inspections (discussed in Chapter 5), are generally conducted on a fee basis, with the fee being sufficient to offset the cost of inspection. Normally the cost of the inspection service is passed on to the consumer in the electrical contractor's charges for the job. Building and electrical inspections have another feature in common—the necessity for close surveillance at the time of construction to achieve satisfactory code compliance. A delay of a day or two in making such an inspection can result in overlooking a serious condition from a safety standpoint. The enclosure that might have been completed during that delay might make proper inspection impossible.

In most jurisdictions, enforcement of the electrical code is at the point of the provision of the electrical service. Most jurisdictions require full code compliance prior to the connection of service to a building from the public utility or power company. This procedure is quite effective and can also be used as an enforcement lever in the inspection of existing electrical installations.

A major problem in electrical code enforcement is the home wiring done by people untrained in the finer points of electrical work. The code enforcement agency has little control over such wiring, because normally a do-it-yourself electrician bypasses the formality of obtaining necessary permits.

Another factor in electrical code enforcement is the requirement for a high degree of technical competency, including licensure procedures, for electricians at various levels of expertise. This is necessary because of the nature of the work; much of the potentially dangerous wiring is hidden on completion of the structure.

HOUSING CODES

Another type of code that has a bearing on fire prevention is the housing code. In recent years, most larger communities in the United States have adopted housing codes in an effort to assure citizens of adequate housing facilities, especially those in rental properties.

Housing codes as a rule include a number of regulations relating to fire prevention and life safety. For example, means of egress and heating appliances are covered by code provisions. Space heaters are not permitted under model housing codes. The right of entry for inspection under housing code authority is generally broader than that for fire prevention inspectors. This provides the housing inspector with an instrument for upgrading life safety in many occupancies that may be inspected by fire department personnel only on invitation.

Because of the personal nature of housing code enforcement, many jurisdictions have been reluctant to permit the implementation of such a code. Many citizens feel that obligatory inspections of individual homes under a housing code represent an unwarranted intrusion on privacy.

PLUMBING, HEATING, AND AIR-CONDITIONING CODES

Mechanical codes, including heating, ventilation, and air-conditioning as well as plumbing codes, will be treated together because they are usually enforced by the same agency in a community. Plumbing codes have a bearing on fire protection, especially when requirements related to automatic sprinkler protection are included as a part of the plumbing code. The mechanical or plumbing inspector may be responsible for seeing that automatic sprinkler equipment is properly installed. The inspector may also have responsibility for checking underground connections relating to fire protection services and for checking hydrants, standpipes, and other features relating to fire protection water supplies.

Mechanical codes, which cover heating and air-conditioning, have a major effect on fire prevention in that they usually include coverage on duct work and other matters related to distribution through heating and air-conditioning systems. Improper installation has resulted in the spread of smoke and fire; in fact, this sometimes occurs even with proper installation. Dampers, floor penetrations, smoke detectors, and duct materials are among fire safety considerations in these installations.

Other codes having a bearing on structural fire safety include codes pertaining to elevators; elimination of architectural barriers to the physically challenged; energy conservation; gas fitting; historic preservation; manufactured buildings; mobile homes; and pressure vessels and boilers. Many states and communities have not adopted all of these codes. Where adopted, responsibility for enforcement is varied. The Americans with Disabilities Act, signed into law in 1990, addresses requirements which public buildings and workplaces must meet to provide access for the mobility impaired.

Historic preservation codes are enacted to encourage the retention of historically significant structures in a community. These codes may permit less stringent fire protection requirements or alternative protection provisions for existing buildings than for new ones. It is impossible to renovate some historic buildings in such a manner as to be in full compliance with codes for new structures.

FORESTRY CODES

Usually administered at the state level, forestry codes and regulations have a definite effect on fire prevention. These regulations are generally in effect only within a given number of feet of wooded or forested areas, and in many states they are not applicable within incorporated municipalities. Regulations often prohibit open burning during periods of low humidity. They may also require safety precautions in connection with the use of matches and smoking materials.

Forestry laws are generally comprehensive and prohibit a wide variety of unsafe acts in forest lands. Among acts generally prohibited are leaving campfires unattended, open burning without a permit during certain seasons, operating a vehicle without a muffler in forest lands, and dropping or throwing burning matches, lighted cigarettes, or other burning materials in or near woodlands. The operation of railroads in woodlands is also closely regulated because of the danger of fires as a result of sparks. Adjacent lands are required to be cleared to reduce this danger.

In many states people who set fires on property they own and permit the fires to escape and become uncontrolled are responsible for payment of all costs for fighting the forest fires. The fires may have been set originally to destroy debris or rubbish.

Forestry regulations are usually well-known, at least by people who work in the forested lands and by those who frequent them for recreational purposes. Because of the large landmasses covered by such regulations, enforcement is often difficult and may be compared to enforcement of maximum speed limits on major interstate highways. It is most uncommon for an enforcement official to see a violation being committed, and enforcement generally comes about through individual complaints with occasional spotting of dangerous practices during routine patrol.

Several counties, primarily in the western states, have adopted provisions that require minimum clearances between woodlands and structures built for habitation in forested areas. These requirements have proven to be valuable in preventing structural damage in wildland fires.

ENVIRONMENTAL REGULATIONS

Air pollution regulations, which are normally enforced by health and environmental agencies, have a bearing on fire safety. Regulations usually require permits for all open burnings; they also impose severe restrictions on locations at which open burning may be carried out and specify narrow time frames for necessary burning.

These regulations have a positive effect in the fire prevention field through the requirement of burning permits and consequent patrol activities designed to reduce nonapproved burnings. Air pollution regulations may preclude the burning of buildings by local fire departments for hazard abatement and training purposes. Normally, some provision is made to permit bona fide training activities to take place; however, this is not always the case.

Another way in which environmental regulations may conflict with fire prevention is by creating the need for storage of combustibles awaiting pickup. Wastepaper and other trash that would normally be burned is held for a period of time awaiting pickup by trash trucks for delivery to the proper location. During this storage time the hazard within the facility may be increased as a result of holding these combustibles.

Again, as in the case of housing code enforcement, it is highly desirable for the fire prevention inspector to maintain close liaison with personnel in the air quality control field. Many misunderstandings can be resolved by this cooperation, particularly with respect to requirements for burning permits. A permit from air quality control is often required in addition to the fire department permit in communities that operate under a model fire prevention code. Forestry agencies may also require permits, so it is conceivable that a person might have to get three different permits just to burn one small pile of trash in the yard. To eliminate this kind of overlap, some communities have combined air quality control, forestry, and fire prevention bureau permit requirements.

Public support for air quality control measures has been strong, which works to the advantage of efforts to achieve 100-percent control through issuance of permits. A fairly high percentage of burnings are being reported to the air quality control people because of public interest and support. Many of these calls are in the form of complaints by neighbors who take exception to burnings occurring in their neighborhood.

Water resource protection regulations have an effect on fire prevention in that oil pollution is almost always covered. There should also be joint efforts between water resources and fire service personnel in the handling of leaks, spills, and other incidents where waterways or water sources may be contaminated.

CODE ADOPTION

A state fire prevention code is in effect in most states. These codes are promulgated by the state fire marshal, the insurance commissioner, the state fire board, or some other agency granted the power to promulgate regulations. Usually, a public hearing is required so that the public, special interest groups, and fire service personnel have the opportunity to address the fire marshal or fire prevention commission regarding possible problems or advantages they see in connection with implementation of the requirements.

Promulgating regulations by commission or board is much better in many aspects than initiation of requirements through statutory provisions. Statutory requirements are far less flexible and may be misleading to individuals who attempt to interpret the limited amount of wording in the statute.

The fire marshal's office may depend heavily on the regulations of other state agencies. For example, department of health regulations may include specific fire requirements that are actually enforced by the state fire prevention bureau or fire marshal's office.

◆ ENFORCEMENT

While the word *enforcement* is an appropriate one, some have suggested that it is a bit harsh and feel that *compliance* is better since that is the ultimate goal and may be achieved by either education or enforcement. It is likely that many early transgressors of fire prevention rules were persuaded to change their ways by seeing enforcement measures inflicted on their neighbors.

Fire and building codes represent the "Engineering" phase of the equation. But without enforcement, or compliance as previously noted, engineering aspects might be forgotten. Somehow the architect and builder must know to install automatic sprinklers; the occupant of a low-rent home must know that a working smoke alarm is needed.

A statement by Mary L. Corso, Washington State Fire Marshal, contained in the *Proceedings of the Solutions 2000* symposium, points out the problems encountered in bringing about fire safety enforcement:

> Who is responsible for fire safety enforcement? Effective enforcement requires a collaborative effort between enforcement agents, those who are responsible for providing fire protection, the community, and government regulators. The community, largely the fire service, is responsible for addressing the fire safety needs of the populations residing within. Providers are responsible for knowing codes and regulations and understanding their benefits. Government regulators are responsible for setting standards meant to ensure a safe environment. Lastly, it is the responsibility of the enforcement community to see that these standards are being followed. However, it is important that each group understands the unique issues of this vulnerable population.[19]

Marshal Corso was referring to young children (under 5) and older adults (over 65) by use of the term *vulnerable*. This symposium addressed fire safety issues for these population categories.

FALSE ALARMS

"False alarms, repeated frequently, may perhaps lead some of the men to move more slowly under the impression that every alarm given is false, and merely intended for exercise." This warning, found adjacent to the fire pump on the USS *Constellation* built in 1854, is as applicable today as it was to sailors on board that wooden hulled vessel. False alarms, once thought of as primarily malicious acts, have increased in many jurisdictions because of the increased sensitivity of many smoke alarms. Although not directly a fire problem per se, these calls can have a delaying effect on the response of fire fighting equipment to a real emergency.

◆ **SUMMARY**

Sound fire prevention practices that produce better fire safety are the product of the three Es—Education, Engineering, and Enforcement. Education requires delivering the appropriate safety message to the affected or at risk group. Education is effective but is dependent upon the commitment of local government or local fire service. Engineering involves fixed system and construction standards as a means of preventing and/or controlling fires. It also promotes life safety not only through prevention but also by providing adequate means of egress. Enforcement means just that, a means to ensure those laws, ordinances, codes, and standards are followed. Whether intentional or not, many buildings would not be built to appropriate standards without adequate enforcement. A community with all three elements—education, engineering, and enforcement—is a safer community.

Review Questions

1. The three Es of fire prevention are
   a. Education, Engineering, Exercise
   b. Education, Experience, Enforcement
   c. Exercise, Engineering, Enforcement
   d. Education, Engineering, Enforcement
2. In 2002, how many U.S. cities had no recognized fire prevention or code enforcement program?
   a. 0
   b. 1
   c. 2
   d. 3
   e. 4
3. Between 1977 and 2002, the death toll in house fires was reduced
   a. 17 percent
   b. 29 percent
   c. 41 percent
   d. 57 percent
   e. 79 percent
4. Children under the age of _____ are twice as likely to die from fire than the rest of the population.
   a. 3
   b. 5
   c. 8
   d. 13
   e. 16
5. Fire death rates in rural areas are how much higher than in nonrural areas?
   a. 12 percent
   b. 24 percent
   c. 36 percent
   d. 48 percent
   e. 60 percent

6. Which statement is false?
   a. Building codes address fire safety requirements
   b. Fire prevention codes cannot be applied retroactively
   c. Building and fire prevention codes may overlap
   d. Nonlife safety elements of a building or fire prevention code are generally not applied retrospectively

7. The two model fire prevention codes are
   a. NFPA Uniform Fire Code and ICC Fire Prevention Code
   b. BOCA and Uniform Codes
   c. Southern and BOCA Codes
   d. NFPA Uniform Fire and Southern Code

8. Noncombustible construction is typically
   a. Type I
   b. Type III
   c. Type V
   d. Type I and II
   e. Types III and IV

9. Fire prevention codes can be classified as
   a. prescriptive
   b. performance
   c. both a and b
   d. none of the above

10. Codes other than building and fire prevention that affect fire safety include
    a. zoning
    b. electrical
    c. housing
    d. forestry
    e. all of the above

■■

Answers

1. d		6. b	
2. e		7. a	
3. d		8. d	
4. b		9. c	
5. c		10. e	

■■

Notes

1. Powell Evans, comp., *Official Record of the First American National Fire Prevention Convention* (Philadelphia: Merchant and Evans Co., 1914).
2. Research Triangle Institute, International City Management Association, National Fire Protection Association, *Municipal Fire Service Workbook* (Washington, D.C.: U.S. Government Printing Office, 1977), p. 4.
3. Federal Emergency Management Agency, U.S. Fire Administration, National Fire Protection Association, *A Needs Assessment of the U.S. Fire Service* (2002), p. v.
4. James M. Shannon, "First Word," *NFPA Journal*, National Fire Protection Association (March/April 2004), p. 6.
5. Federal Emergency Management Agency, U.S. Fire Administration, *Babies and Toddlers Fire Death Data Report* (Emmitsburg, Md., 2003).
6. Canadian Association of Fire Chiefs, *Canadian Fire Chief* (Ottawa, Ontario, Summer 2003), p. 29.
7. Federal Emergency Management Agency, U.S. Fire Administration, *A Profile of the Rural Fire Problem in the United States* (1998), p. 3.
8. American Burn Association, *2004 Burn Incident Fact Sheet*, undated.
9. "6 die in Chicago office fire," *The Gainesville* (Fla.), *Sun*, October 18, 2003.
10. *NFPA. Wildfire Notes and News* (September 2003), p. 1.

11. *Fire Suppression Rating Schedule* (New York: Insurance Services Office, 1980).
12. Vincent DiMase, *Building Codes as an Aid to Fire Protection and Fire Prevention* (Providence, R.I.: 1975).
13. Personal communication to the author from the International Code Council.
14. Federal Emergency Management Agency, U.S. Fire Administration, *Evaluating Performance-Based Designs, Student Manual* (Emmitsburg, Md., 2003), p. SM1-4.
15. Ibid., p. SM1-9.
16. Ibid., p. SM1-8.
17. Ibid., p. SM1-12.
18. Ibid., p. SM1-7.
19. Federal Emergency Management Agency, U.S. Fire Administration, *Report for Beyond Solutions, 2000,* undated, p. 25.

CHAPTER 3

Public Fire and Life Safety Education Programs

F ire and life safety education is receiving more recognition than ever before by the fire services of North America. Although public fire education is now a well-established practice in a great many communities in the United States and Canada, it is not yet universal, and more effort must be directed toward this goal.

America Burning, the report of the National Commission on Fire Prevention and Control, recognized the importance of fire safety education. It stated, "Among the many measures that can be taken to reduce fire losses, perhaps none is more important than educating people about fire. Americans must be made aware of the magnitude of fire's toll and its threat to them personally. They must know how to minimize the risk of fire in their daily surroundings. They must know how to cope with fire, quickly and effectively, once it has started."[1]

To get public fire safety education programs under way, there must be a spark plug in the fire department, someone who has attended programs or had experiences that stimulated an interest in fire prevention. This individual can encourage others in the department to recognize the potential of fire prevention. The department's administration must also have a compelling desire to initiate the programs. However it may come about, acknowledgment of the important role of fire safety education among the fire department's responsibilities is a mark of the modern fire service, whether the department be career, volunteer, call, or part paid.

◆ SCOPE OF FIRE AND LIFE SAFETY EDUCATION PROGRAMS

Fire prevention education is primarily the dissemination of information relating to fire hazards and fire causes in the hope that the public will take proper precautions against fire. To conduct educational programs, fire department personnel should be trained in public speaking and staging demonstrations. In fact, several fire departments have incorporated formal training in public speaking in their recruit training programs.

The term *fire prevention education* generally includes fire reaction–related training as well. Chapter 8 is devoted to factors relating to fire reaction; however, a great number of fire prevention education programs include this subject. The term *public fire safety education* includes both activities—fire prevention and fire reaction. Burn awareness education is also included.

Many fire safety education programs have been expanded to include injury prevention. The inclusion of other safety subjects must be carefully considered lest the original fire safety message be lost. While all contribute to the welfare of the community, they must be administered in doses that can be easily absorbed by the particular audience. Widespread inclusion of emergency medical service as a fire department function supports the injury prevention concept. The term *fire and life safety education* encompasses this expansion of goals.

Wildfire mitigation has become a part of many fire departments' goals in public education. Often state and national forestry organizations share in this responsibility. With the growing occurrence of wildland–urban interface as a major fire problem, fire prevention agencies have added this dimension to their repertoire of activities. Many jurisdictions have adopted programs similar to that adopted by Colorado Springs.

Colorado Springs, for example, has a Wildland Risk Management Office within its fire department. The fire marshal is a major player in this activity. Much of its program is based upon NFPA's *Fire Wise* initiative, but adapted to its city's needs. The program involves an evaluation of hazards for homes located in wildland areas with community educational programs on abatement measures that may be initiated by the homeowner. The program limits roofing materials to those that meet fire safety standards, and it also includes an ordinance giving the fire chief the authority to order evacuations in extreme danger situations.[2]

STEPS IN PUBLIC FIRE AND LIFE SAFETY EDUCATION PLANNING

As suggested by the revised version of the manual *Public Fire Education Planning, a Five Step Process,* the steps involved in developing and operating a community risk fire education program are listed here:

Step 1: *Conduct Community Analysis:* A community analysis is a process that identifies fire and life safety problems and the demographic characteristics of those at risk in a community.
Step 2: *Develop Community Partnerships:* A community partner is a person, group, or organization willing to join forces and address a community risk. The most effective risk reduction efforts are those that involve the community in the planning and solution process.
Step 3: *Create an Intervention Strategy:* An intervention strategy is the beginning of the detailed work necessary for the development of a successful fire or life safety risk reduction process. The most successful risk reduction efforts involve combined prevention interventions (Education, Engineering, and Enforcement).
Step 4: *Implement the Strategy:* Implementing the strategy involves testing the interventions and then putting the plan into action in the community. It is essential that the implementation is well coordinated and sequenced appropriately. Implementation occurs when the intervention strategy is put in place and the implementation plan schedules are followed.
Step 5: *Evaluate the Results:* The primary goal of the evaluation process is to demonstrate that the risk reduction efforts are reaching target populations, have the planned impact, and are demonstratively reducing loss. The evaluation plan measures performance on several levels—outcome, impact, and process objectives.[3]

Effective educational program materials should be designed and developed. Suitable materials may be available from national fire safety organizations. In the design of material, the first goal is to determine message content. Messages should be directed to specific hazards. They should appeal to positive motives and not be

FIGURE 3.1 ◆ Fire hazard houses assist public fire safety educators in getting their messages across. *(Mead Westvaco, Luke, Maryland)*

threatening. The messages should show the context of the problem and desired behavior. The format, whether it be a wall chart, video, PowerPoint, or folder, should be matched to the message, audience, and resources.

The planner should determine the appropriate delivery time by specifying the target groups and finding out when those audiences will be most receptive to fire and life safety messages. Messages should be scheduled for maximum effect. *Message,* as used here, includes any type of dissemination of information, whether it be through home inspections, media releases, television, or public fire safety demonstrations.

The program package should be designed and subsequently presented to a sample audience. If found to be effective, the program is then implemented. Material is purchased or produced and distributed, instructors are trained and scheduled, and audience participation and cooperation is obtained.

Programs should be constantly monitored and modified on the basis of the monitoring review. They should be evaluated for effectiveness through review of loss data and educational data. Telephone polls have proven valuable in obtaining such information. As important, however, is the evaluation of program impact. The results should be favorable if the proper steps have been followed in program development.

◆ HOME INSPECTION PROGRAM

The home inspection program is primarily devoted to fire prevention education. This program probably began in the United States in May 1912 in Cincinnati, Ohio, when the fire department started a comprehensive home and business inspection program.

By May 1913, the department had experienced a 60-percent reduction in fires. These inspections were made by personnel detailed from each of the 45 companies in the department. There were 80,000 structures in Cincinnati at the time, and the program was aimed at reaching each of these on an annual basis.[4] Radio did not exist at the time, and fire apparatus was primarily horse-drawn.

In January 1913, the Cleveland, Ohio, Fire Department started deploying fire personnel from the 35 steamer pumper companies and 13 truck companies to make inspections. During the first nine months of 1913, 249,000 inspections were made, which resulted in the removal of 29,000 loads of rubbish, among other rewards.[5] The Portland, Oregon, Fire Department started home inspections in 1913.

It is interesting to note that Mr. Powell Evans, in a talk to a national fire conference in Philadelphia in October 1913, stated:

> Fire departments, as a rule, all over the country, up to this time have been selected and managed on old "rule of thumb" lines and insufficiently instructed and brought up to modern methods. After emphasizing the need for fire fighting schools, we pass on to our particular subject—the use of active firemen for fire prevention purposes; which provides better occupation for the men, familiarizes them with the property under their control and enables them to perform their hard duty in emergencies with greater speed and safety than otherwise. This system has been employed in a few cities as long as 15 years and is rapidly spreading. It has worked to advantage wherever used. . . .[6]

In 1921, the Wilmington, Delaware, Bureau of Fire started a home inspection program, and Providence, Rhode Island, began one in 1930.

During the first week of the program in Providence, which was during Fire Prevention Week, the inspections resulted in the removal of 1,680 tons of combustible material that had been stored in attics and cellars for years. Included were 2,800 discarded mattresses and 1,000 dry Christmas trees. These trees had been stored since the previous Christmas season.

VOLUNTARY BASIS

As noted earlier, the home inspection program is primarily a fire prevention education endeavor. Although the word *inspection* is used, there is no legal backing for the program in most jurisdictions. Entry into residences is on a voluntary basis, and if at any time occupants look on the program as being mandatory and believe they are being forced to make changes, there is a good chance that it will collapse. Personnel assigned to home inspection duties should be fully apprised of the fact that the program is entirely voluntary.

TRAINING

Fire service personnel must be properly prepared before embarking on home inspection duties. The training program must encompass details of items to be checked, including potential hazards and smoke alarms; the importance of personal neatness and courtesy; and public relations aspects of the job. Successful programs have been carried out in volunteer as well as career and part-paid fire departments.

ADVANCE PUBLICITY

To prepare the public to accept home inspections, the program must receive a considerable amount of advance publicity. This publicity should describe the purposes of the program and mention inspection, with particular emphasis on the fact that the program is strictly a voluntary one.

Planning should encompass schedules for areas to be covered each day, and definite routes should be planned to schedule personnel in a way that makes most effective use of their time. There is no reason why nonuniformed personnel, as may be the case with some volunteer departments, cannot participate. They can be provided with identification badges, and they should wear caps or other identifying clothing.

◆ FIRE PREVENTION EDUCATION THROUGH CIVIC ORGANIZATIONS

One of the best means of promoting fire safety education is through civic organizations. Organizations such as the Rotary, Lions, Civitan, and Kiwanis are interested in projects that may be of assistance to the community. Most of these organizations have a wide range of community interests represented among their members. This variety of membership is helpful to the fire preventionist who is attempting to solicit the support of the organization. The contacts that can be made through a civic organization can result in garnering of much support for the fire department in the community.

CHAMBERS OF COMMERCE

Another civic organization that can play a part in fire prevention is the chamber of commerce. Local chambers of commerce are usually encouraged to cooperate in fire prevention efforts in the communities they serve. Chambers of commerce may hold meetings in the community to develop greater interest in fire prevention.

OTHER LOCAL GROUPS

Local parent–teacher organizations can also assist in promoting fire prevention programs in the schools as well as in the community. The interests of such organizations are primarily in school fire prevention; however, the groups may wish to have fire and life safety education programs as part of their regular meetings. They can be of considerable assistance in supporting the correction of fire safety deficiencies in the schools.

As a means of promoting fire prevention, some communities have developed programs to appeal to church congregations. These have emphasized fire prevention measures in the church but have also included suggestions that can be taken into the home. Some communities have even staged fire drills in churches in the interest of improving fire evacuation performance. Fire safety messages may be printed in church bulletins.

Fire buff organizations have been instrumental in supporting fire prevention activities in some cities. Members of these clubs represent a cross section of the community and share a common bond of strong support for local fire department activities. Members can be of great assistance in promoting fire prevention activities, in serving coffee for fire prevention open houses, and in raising funds to purchase fire prevention material that might not otherwise be available.

NON-ENGLISH-SPEAKING GROUPS

In many communities special efforts must be made to ensure that the audience can understand the fire education message being conveyed. Some communities contain large numbers of individuals who are unable to speak or read English. This means that the fire prevention bureau chief must see that programs are designed to reach these individuals. It is helpful to have fire department members who are able to speak the appropriate language. Means must also be available to communicate a fire safety message to those who have difficulty with sight and hearing or have other limiting disabilities.

◆ FIRE SAFETY CLINICS AND SEMINARS

The fire safety clinic or workshop has been a valuable aid in fire prevention efforts in some communities. Under this concept, representatives of industry and institutions are encouraged to attend a fire safety program during which all phases of fire prevention and reaction are discussed. Some of these programs include outside demonstrations that are most helpful in creating an interest in fire safety.

Seminars of this type are especially valuable if they are aimed at one particular type of audience. If a general audience is anticipated, it is difficult to develop a meaningful program to last for a day. Seminars may be designed primarily for department store personnel, apartment house operators, or nursing home and hospital personnel, as examples.

◆ COMMUNITY EVENTS

Fairs and other organized activities have given a number of communities an opportunity to promote fire prevention. These activities present a chance to set up displays and booths for public informational purposes. They give the fire department an opportunity to have a great deal of personal contact with the general public. Participation at a fair may involve staging displays of fire fighting equipment and fire safety demonstrations.

Inviting citizens to visit a home that has experienced a serious fire provides a meaningful safety message. Several cities have held open houses at recently burned dwellings to let the public see the ravages of fire. This can be done only with the permission of the homeowner; however, many victims welcome the opportunity to help mold positive attitudes toward fire safety. Fire department personnel may prepare the dwelling by placing signs to indicate point of origin, fire spread characteristics, and other salient features of the fire. Personnel should also be on hand to answer questions. Safety of the visitors is a primary concern.

The staging of the "Burning of the Greens" is another activity that can be most helpful from a fire prevention standpoint. These programs normally include a fire prevention message as well as a message on the potential dangers of discarded Christmas trees that are being destroyed. Air pollution regulations have restricted this measure of promotion.

"LEARN NOT TO BURN" AND RISK WATCH PROGRAMS

The "Learn Not to Burn" program of the National Fire Protection Association has been credited with saving many lives. The program entails several segments: public service television announcements, in which brief messages are given by a well-known personality; a "Learn Not to Burn" curriculum designed to be used in preschool through third grade; a high school program entitled "Fire Safety for the Rest of Your Life"; Fire Prevention Week, the nationwide fire prevention observance sponsored by the National Fire Protection Association since 1922; "Sparky," the symbol of fire safety for children; and the network of fire and life safety education representatives who assist in implementing the preceding programs.

The "Learn Not to Burn" curriculum is designed to be taught by teachers in the school system with the assistance of fire service personnel. In some cases, teachers have been unable to adjust their schedules to include "Learn Not to Burn," despite the fact that the local school district desires its inclusion in the overall school curriculum. Volunteer instructors, including firefighters, may fill this role. Another popular offering from NEPA is the "Remembering When" program for older adults which teaches fire and fall prevention.

Risk Watch, another National Fire Protection Association program, is a comprehensive injury prevention curriculum for children in preschool through grade eight. It is designed to target the major unintentional injuries which threaten children ages 14 and under, including fire and burn prevention, and it includes a natural disaster module. Risk Watch has been adopted statewide in several states and has proven to be quite popular.

◆ FIRE AND LIFE SAFETY EDUCATION IN THE SCHOOLS

Opportunities to present fire prevention and reaction messages in schools are limitless. Through the years there have been better chances of getting a fire prevention message to youth groups than to any other segment of the population. The school system must in all cases be consulted before planning any program that is addressed to school personnel or students. The school system, of course, has primary responsibility for educating students, and this responsibility can in no way be ignored.

A common school fire prevention effort is an appearance before school assemblies. A fire prevention officer can address a school gathering without having to give the talk time and time again in individual classrooms. An appearance before a large group has the disadvantage of limiting personal contact and subsequent questions and discussion. A discussion session can be much more successfully conducted with smaller classroom groupings dealing with age-appropriate material. The opportunity to appear before the entire student body may occur less frequently than opportunities to appear in individual classrooms. The fire department may be able to arrange both types of presentations.

With pre- and posttesting methods, it is possible to assess the relative value of the approaches. There is no question that a consistent, positive program of active learning experiences, reinforced over time, is much more beneficial than a once-a-year "tell them everything you know" presentation.

Although not directly related to fire prevention education for students, a great deal can be said for the inspection of schools as a fire safety measure. The presence of fire department personnel in the school is a desirable feature from a fire protection standpoint.

Home inspection forms have proven to be a highly effective means of getting across the message of fire safety. Students are given these blank forms to take home and are encouraged to return the completed forms. Instructions for completing the form must be given, as well as careful feedback on the responses to ensure accuracy. Most students are highly motivated in getting the job done. Parents usually assist with filling in the forms. If such a form is used, thought must be given to the level of understanding and ability of the student. There is no value in using a form that cannot be readily understood by the child filling it in or by the parent. Home escape plans and smoke alarm information may also be sent home with students.

FIRE AND LIFE SAFETY STUDIES IN THE CLASSROOM

The fire prevention program for public, private, and parochial schools should include training on fire safety. This training might include information on outdoor fire hazards such as the burning of leaves and indoor fire hazards such as heating appliances, electrical hazards, kitchen equipment, and housekeeping.

The training program should strive to develop a knowledge of fire and life safety measures that might be employed by young people. Fire prevention personnel have sometimes misdirected training efforts by not considering the learning and comprehension abilities of the students being taught. There may be a problem whereby a training program sparks an interest in trying out fire in certain age groups. Psychologists have indicated that many people have a natural curiosity about fire, especially younger children. Fire prevention efforts should not encourage this interest.

It is possible at the junior high or middle school level to include fire prevention in other courses. History classes, for example, could cover some of our national major fires to develop a greater realization of fire as a problem. Likewise, civics classes might include discussion of the fire department as a part of the city government, with emphasis on the cost of fire protection in the community as a part of the municipal budget. Factors relating to the complete impact of fire should also be included. The role of fire insurance might be mentioned as well. Likewise, fire safety might be included in the study of economics. The economic impact of fire on society could well be a subject of interest in an economics class.

Fire prevention may be taught in science classes. Experiments in the chemistry of fire could be of considerable help in teaching fire safety at the middle school and high school levels.

Shop subjects can also include fire prevention. There are many opportunities for the inclusion of fire prevention in shop classes because they often include the use of flammable liquids as well as the storage of rags and other combustible materials.

The overall program for fire education in the school system should include some contact at each grade level. Unfortunately, personnel limitations and other factors often limit the activities to one or two grades.

Some states have had mandatory fire education programs for a number of years. Ohio passed a law in 1908 requiring "that every teacher in every public, private or parochial school shall devote not less than thirty minutes in each month during which the school is in session, to instruction of pupils between the age of six and fourteen years, in fire dangers." The state fire marshal was required to prepare a textbook to be published by the state and provided to each teacher.[7]

At least 35 states require fire safety education in the schools. Implementation of required programs is not well enforced in some states, however, and many feel that locally developed nonmandatory programs are more effective.

COORDINATING FOR SUCCESS

Close coordination with school faculty is necessary for the development of successful fire and life safety education programs in the schools. Many school systems have a standing policy against including material in the curriculum that has been developed outside the system. A spirit of cooperation with appropriate personnel in the school system may result in modifying such a position to provide at least some fire prevention coverage where no material has been developed in the education system.

Another problem in school fire prevention programs has been with the distribution of commercial fire safety material in the schools. Fire departments in a number of communities have asked the school system to permit the distribution of fire prevention literature published by insurance companies, candy companies, or other commercial enterprises. Although due to budgetary limitations the fire department may have no choice but to use this material, the school authorities in some communities do not feel that they can permit the distribution of fire safety material advertising an individual commercial organization.

Two locally developed programs are described here.

Fire and Life Safety Educators are becoming more creative in developing programs and ways to get safety information into preschools and elementary schools. Some of this comes from the challenge to have schools approve time away from the curriculum for safety. It seems fire departments continually compete with the many demands placed on teachers and schools.

Two such programs were developed by the Lakeland, Florida, Fire Department. The first program, "Fire Safety Traveling Trunks," provides preschool and kindergarten teachers with all of the materials needed to teach fire safety in the classroom. The trunks include items for the various centers that are normally set up in preschool and kindergarten classrooms. The trunks include the NFPA "Learn Not to Burn" preschool curriculum and such items as child-sized gear for the dress-up center; matching cards, sequencing cards, and counting cards for the math center; an interactive fire safety computer program for the computer center; and audiotapes of fire safety songs for the music center. The trunks also include items to be used during circle time such as a black sheet to practice *crawl low under smoke* and felt flames for practicing *stop, drop, and roll*. The Traveling Trunks were developed to enhance the fire safety education, not replace the educational programs presented in the schools and centers. However, the motivating factor behind the trunks was to get the teachers to help teach fire safety in the classroom. The number of preschool and kindergarten classes in this area continues to grow and it is increasingly difficult to present programs at every center every year. Teachers utilize the materials in the trunks that are in their classroom for two weeks and the materials are user friendly.

"Story Time Is Safety Time" is a second program developed for preschool children to fit into "Circle Time" or "Story Time" in the classroom. The preschool story time includes stories, songs, finger plays, and action rhymes and utilizes a felt board. These added elements enable the program to be correlated to a variety of the local/state standards. Story time can be used for various levels as long as the elements of the program are age appropriate. Story time can also be presented at other locations that host story times, such as the local library, children's museums, and bookstores.[8]

JUNIOR FIRE MARSHAL PROGRAM

The Junior Fire Marshal program has been effective in many sections of the United States. Both English and Spanish versions are available. In some areas this program has been sponsored by a local insurance organization, with the field work done by fire department personnel. In most localities the material has been supplied by the school district or fire service, and all field operations have likewise been under that agency. Membership in the Junior Fire Marshal program is usually obtained by filling in a home fire prevention form or by taking some other step that aids the cause of fire prevention. School students enrolled in the program are usually given certificates of fire safety knowledge.

Sponsored by the ITT Hartford Group, the Junior Fire Marshal program is designed for children in kindergarten through third grade. It focuses on fire prevention tasks through the use of teacher guides, films, filmstrips, and audiocassettes. Fire safety in apartments as well as single-family homes is addressed.

◆ HOSPITAL PROGRAMS

Many fire departments have become involved in the SAFE KIDS Campaign. This organization, through state and local chapters, promotes bike helmets and bike safety, child vehicle occupant protection, and general injury prevention as well as burn prevention and fire safety. The campaign is a program of the Children's National Medical Center in Washington, D.C.

Burn prevention programs operated by hospital burn units may be successfully combined with fire department prevention programs. This combination is used successfully in school programs staged by hospital burn units and the local fire departments. Burn awareness activities of the American Burn Association may be implemented by fire services as well.

A number of burn centers are active in public fire safety education. Statewide burn foundations also sponsor similar programs. An outstanding example is the Mississippi Firefighters Memorial Burn Center in Greenville. This center was established through the efforts of the state's firefighters. They were concerned about the dearth of burn treatment facilities in the area, which has a high per capita burn rate. This direct link between the fire service and burn treatment brings about a greater appreciation of burn prevention needs by the public.

◆ SCOUT GROUPS

A number of fire departments have had considerable success working with Scout organizations. This work includes helping Scouts earn merit badges, making appearances at Scout meetings, giving demonstrations before Scout gatherings, and arranging visits to fire stations by Scout groups. In addition, Scouts may participate in wastepaper drives, salvage drives, and other efforts that result in improved fire prevention practices.

Scout groups have also assisted with baby-sitter training programs. The baby-sitter training programs normally given by fire departments can be readily adapted to Scout groups. Programs can also be adapted to 4-H groups and other youth groups. Such programs may include specific information for farm fire safety problems.

◆ WILDLAND FIRE PREVENTION

As previously noted, wildland–urban interface fire abatement has become a part of the responsibilities of many fire departments. A team approach is used in Texas to assist smaller cities and counties. These wildland fire prevention teams assist local agencies in prevention of human-caused fires and minimize property losses through owner education. These teams work together to:

- complete fire risk assessments
- determine the severity of the fire situation
- facilitate community awareness and education in fire prevention (including prescribed burning)
- coordinate interagency fire restrictions and closures
- coordinate fire prevention efforts with the public, target groups, agencies, and elected officials
- promote public and personal responsibility regarding fire prevention in the wildland–urban intermix
- plan for fire protection[9]

◆ PUBLIC SAFETY COMBINED PROGRAMS

Public interest in crime prevention is generally greater than that in fire prevention. For this reason, some jurisdictions have combined crime prevention and fire prevention education programs. Cities with public safety departments have found this to be a desirable concept. In addition, some communities recognize that the police department is often more apt to be invited to deliver a crime prevention message, and therefore some jurisdictions have police and fire personnel participate in the same session. Usually, the program starts with crime prevention, and then the fire service person concludes with a fire safety message.

An example of a program in which fire safety and other public safety issues are addressed was conducted in North Dakota:

The Fire Marshal's Office joined forces with two other state agencies that needed to get information to the state's senior citizens, the Bureau of Criminal Investigation (BCI) and the Consumer Fraud Division. BCI had a need to address crime affecting the elderly and Consumer Fraud had a need to make elderly consumers aware of the great number of "rip-offs" involving senior citizens. The three state agencies put together a program called "The Protection Connection" and made presentations across the state. The vast majority of presentations were made at senior citizen centers. This approach seemed to work very well.

The fire safety portion of the program was divided into two segments. The first segment dealt with fire prevention and included the topics of: electrical safety, kitchen fire safety, fire safety in the living room and bedroom, and general fire safety. The second portion of the program focused on surviving a fire and steps to take when a fire occurs. Statistics and fire facts were used to emphasize the seriousness of the problem. The next portion of the program dealt with smoke detectors including testing, changing batteries, and proper placement. The next topic was the need to develop a fire escape plan. Emphasis was placed on getting out fast when a fire occurs, having two ways out, having a meeting place outside the home, and a plan to call the fire department. "Stop, Drop, and Roll" was also a part of the presentation. At the end of the program, time was allowed for questions and answers. On many occasions this segment turned into a kind of "show and tell" with the audience sharing experiences they had with fire. We concluded the program by distributing several hand-outs including "Fire Safety Tips for Older Adults," published by NFPA.[10]

◆ **PUBLICITY PROGRAMS**

A variety of programs can be devised to capture the attention of the community and to publicize the theme of fire prevention. Following are some ideas that have been used by fire departments in the United States.

CONTESTS

Poster contests have been widely used in the fire prevention field for many years. These contests are generally conducted by the local fire department or fire service association in cooperation with community schools. The contest should develop an interest in fire prevention and not alienate school officials from the fire service. Some school officials believe that poster contests are inappropriate as a means of achieving fire prevention because the artwork in these contests may not be in keeping with educational programs. Any fire department considering a poster contest should first check with school system officials and obtain their approval before proceeding. Entries must be checked for technical accuracy. The posters may be displayed in the city hall or library.

PROMOTIONAL AIDS AND ACTIVITIES

Place mats bearing a fire prevention message have also proven to be effective. Diners will often read the wording on a place mat while waiting to be served. The use of table tents with printed messages can also be helpful.

Favors such as rulers, pencils, fire helmets, and stickers can be helpful in promoting fire safety. These favors are usually relatively inexpensive and are appreciated by children as well as adults. They may be passed out at public events as prizes for answering fire prevention questions accurately. Favors should not present a safety hazard to recipients.

The hydrant collar is another fire prevention message-bearer used in some communities. Talking hydrants, which are operated by remote control, have proven to be very effective in imparting fire safety messages to young children.

Placards on safety islands, lamp posts, and other city property can be helpful in fire prevention efforts. Posters are also used on fire department vehicles, municipal buses, and municipal garbage trucks. Cities with subway systems have a great opportunity for presenting fire prevention messages through the use of placards.

Milk cartons contain a panel that can be used for rather complete fire prevention messages. Bread wrappers can also be printed with fire prevention messages. Some communities, with the cooperation of local dry cleaners, have taken the message of fire prevention to the public on dry cleaning garment bags.

Outdoor reader-board and billboard fire prevention publicity can be helpful. At least one major grocery chain also uses these boards for fire prevention messages.

Static fire prevention displays have proven successful in some communities. These displays may be placed in store or bank windows or in other locations where they can be readily observed by the public.

A number of cities have opened permanent facilities in which citizens can learn fire prevention measures as well as fire survival procedures. Milwaukee, Wisconsin's Survive Alive House, a converted recreational field house, provides visitors with an opportunity to practice home fire drills under realistic conditions.

One of the most successful citywide fire safety activities is the Great Louisville (Kentucky) Fire Drill, which has been held annually on Sunday of Fire Prevention Week since 1984. The drill is designed to educate the public in fire safety by giving participants a chance to practice home fire drills and learn to stop, drop, and roll. They also learn about matches and lighter safety and how to report a fire. From 12,000 to 15,000 people attend the event each year. Many other exciting activities take place at the Drill, which rotates among the city's parks so that all neighborhoods have access to the event.

Talking robots are used by a number of fire services. They are especially helpful in attracting small children to fire safety messages in public places such as shopping centers and fairs.

Fire clowns are quite successful in teaching children basic fire safety such as "Stop, Drop, and Roll" and home evacuations. Florida's State Fire College conducts week-long classes to meet the need in this specialized field.

◆ MEDIA PUBLICITY

An important part of any fire prevention undertaking is obtaining publicity through the media. Media releases must be interesting, direct, and to the point. Material that is uninteresting and vague will not receive newspaper, television, or radio publicity. It is necessary for the public information officer, chief, fire marshal, or other individual connected with fire safety publicity to work to develop or locate material that will be accepted and used by the media. The fire prevention officer should visit the media offices to develop proper rapport with their staff. Such a visit may be quite rewarding and can result in long-term cooperation between the media and the fire department.

Any fire safety proclamation that might be developed in cooperation with the mayor's or governor's office should be included in media coverage by the fire department. This includes fire prevention and cleanup week proclamations.

The media should also be encouraged to accurately report fires, and particular attention should be given to including information on contributing factors such as lack of smoke alarms. There is no greater opportunity for disseminating fire safety information than by reporting fires in the community.

Any special events carried out by the fire department should be reported to the media. Many times the department is discouraged when articles do not appear immediately; however, the news media can only use material of a nonurgent nature on occasions when their newsrooms are not already filled with information having a higher priority.

Radio and television stations can contribute a great deal to fire prevention publicity. They can run public service spot announcements. This is among the most effective means of getting out fire prevention messages. Announcements may be developed by the fire prevention bureau locally or by one of the national fire organizations. Locally prepared announcements are usually the most effective; however, some announcements prepared by national organizations are designed to permit wording pertaining to local problems. Messages may address current or seasonal fire problems. The Internet has also proven to be quite valuable in dissemination of fire prevention information.

An example of Internet use for fire prevention is the message offered by the South Jordan, Utah, Fire Department. It announces its baby-sitting course, high school fire science class, and other fire safety programs through this medium.

In many communities, the fire department has been successful in getting television coverage of fire safety demonstrations and other such activities. The fire department will probably not always have prime time available for such promotional programs.

Trailers or buses can be equipped to include a considerable amount of fire safety material for demonstration and may be arranged to allow rapid movement of people through the display. Many communities have outfitted such a unit. In some cases the unit is a surplus bus from the municipal transit company. A talking Sparky can be added to the fire prevention bus or trailer to attract attention to the unit. A number of the trailers are equipped to demonstrate smoke alarm and residential sprinkler operation.

◆ SMOKE ALARM PROGRAMS

There are many smoke alarm distribution programs. In some instances, smoke alarms are installed in residences by firefighters. This program is very easily incorporated into a home inspection program; however, the liability issue should be addressed before program implementation.

In some areas, smoke alarms are distributed to groups and locations that are considered to be high risk. For example, the National SAFE KIDS Coalition is involved in distributing smoke alarms to homes where young children live who are not capable of self-preservation in the event of fire. Some older adults are also considered to be high risk because of impairments that make it difficult or impossible for them to react appropriately in the event of fire. Low-income residents are provided with smoke alarms in many communities by the fire department.

Because smoke alarms provide early warning when fire occurs, this type of program can impact fire loss data in a community. Any smoke alarm program should be ongoing and should address smoke alarm maintenance including battery replacement on an annual basis and replacement of the entire device after ten years' service.

◆ FIRE PREVENTION WEEK

Fire Prevention Week has long been looked on as an opportunity to spotlight fire prevention for all citizens during a concentrated time span. The observance began as Fire Prevention Day through a proclamation by President Woodrow Wilson on October 9, 1920. In 1922, the observance was extended to one week by proclamation of President Warren G. Harding. October 9 was chosen to commemorate the great Chicago fire in 1871.[11]

Fire departments throughout the United States and Canada use Fire Prevention Week as the time for aggressive activities aimed at all segments of the population. Radio, television, and newspaper announcements are extensively used at this time.

◆ VOLUNTEER FIRE DEPARTMENTS

Many volunteer fire departments are providing their citizens with excellent fire safety programs. In that most departments were first organized primarily for fire suppression, these volunteers are really going "the extra mile" to render this service in their communities. Many volunteer fire departments have members with talents and experience in teaching, home repair, and a variety of skills which can be useful in fire prevention activities.

Although it is difficult for most volunteer fire departments to conduct mandatory fire safety inspections, no conflict of interest arises with public education activities. Volunteer fire services usually have individuals who for reasons of health or age can best use their talents in public education work rather than fire suppression. This is true of rural, urban, and suburban volunteer fire departments. In fact, a number of Maryland volunteer departments have excellent fire and life safety education programs.

◆ REVIEW OF SUCCESSFUL PROGRAMS

Proving Public Fire Education Works, a publication of TriData Corporation, describes 77 public education programs in the United States and Canada. Comprehensive community-wide and school programs, specific causes and target groups, juvenile firesetter and smoke alarm programs, as well as national programs and special topics such as those related to wildfire prevention, are covered. Practically all resulted in a reduction in fires, injuries, and fatalities.

The publication includes a summary of factors in common that make programs successful:

- They have "champions" who see the program through and lead its implementation.
- They are situated in departments with magnanimous chiefs who allow their public educators room to be innovative and to seek outside resources.
- They carefully target a particular aspect of fire safety, or strike in force across a broad front over and over, reaching a large percentage of the population.
- Market research in one form or another is used to tailor the programs to their intended audience. Powerful allies are obtained, often in the business community or the education community, to get through the bureaucratic barriers and provide assistance.
- The materials used in the program may not be fancy, but they are clear and in abundant quantity.
- The programs reach a significant percentage of their target audience with public educators often going door to door, literally, or through the media to have a broad impact.
- They often repeat messages over and over just as an ad campaign would.
- The good programs are adaptable, changing goals and materials as the fire problem changes.
- And they often are refined by testing in a small area for a small target population before they are implemented community-wide.[12]

Another TriData Corporation publication, *Reaching the Hard-to-Reach—Techniques from Fire Prevention Programs and Other Disciplines,* contains overviews of methods used to reach these audiences, which include older adults, persons living in poverty, non-English-speaking citizens, and a wide variety of other segments of the population. Sixty-five case studies from all areas of the country are included. These studies reflect on successful programs that may be emulated.[13]

A survey to determine attitudes toward and knowledge of fire safety by the public and firefighters can be quite helpful. Steve Bethke, public educator for the Minneapolis, Minnesota, Fire Department, conducted such a survey. In his canvas of neighborhoods, he found that the average citizen felt that 24 percent of the department's budget is spent on fire prevention when the actual amount is 6 percent. Another question asked whether the respondents felt the fire department was likely to arrive in time to save anyone trapped in a burning building. Eighty-seven percent felt that this was a true statement, indicating a need for greater public education emphasis on this issue as conditions seldom enable such rescues. The Minneapolis survey also found that the firefighters surveyed strongly support the need for public fire education.[14]

◆ FUTURE NEEDS IN FIRE AND LIFE SAFETY EDUCATION

The Wingspread IV Conference, held in Dothan, Alabama, in 1996, brought together fire service leaders under the sponsorship of the International Association of Fire Chiefs Foundation. Earlier conferences had been held every ten years beginning in 1966. Conferees discussed the status of the fire service and areas that need improvement. They made conclusions about the current status of public fire education:

Public fire education has progressed significantly in the past twenty years. Programs now are targeted for specific audiences and specific risks, and are better evaluated. Fire safety messages have been refined. The concept of teaching people how to prevent fire and how to take proper actions in case of fire has been expanded to include teaching people how to prevent many "accidents." This concept is called "all risk prevention," "injury control," "fire and life safety," and/or "multi-hazard-prevention."

Recommendations for continued progress in the field included:

- Develop standards for programs and messages
- Develop more messages about the technology of detection, alarm, and automatic sprinkler systems in residential properties
- Include education of elected and appointed officials
- Use locally based methodologies and initiatives to educate citizens and customers
- Build into programs a method of evaluation to determine whether public education is achieving its goal of behavioral change[15]

◆ SUMMARY

The field of public fire and life safety education offers many opportunities to reduce loss of life and property by fire. The key to success in the field is not always easy to find. Hard work and enthusiastic support are necessary ingredients.

Fire safety education is the dissemination of information relating to fire hazards and fire causes with the hope that the public will get the message and take the proper precautions against fire. It also includes the development of proper fire reaction. Among the avenues used to reach the public is the home inspection program, which apparently had its beginnings in Cincinnati, Ohio, in May 1912. Although the program is called "inspection," participation is voluntary on the part of homeowners.

The fire prevention educator can also work through civic organizations, such as chambers of commerce, parent–teacher associations, clubs, and churches, by giving talks and demonstrations. Community events—parades, fairs, dances, beauty contests, and other organized activities—offer many opportunities to publicize the message of fire prevention and enlist community cooperation.

Fire safety education in the school classroom offers limitless opportunity to instill lifetime attitudes about fire prevention precautions. Students who are taught to be aware of hazards that cause fires and what to do about them can help the cause of fire safety in the home and in the community. School programs can take the form of actual classroom instruction or talks and demonstrations before school assemblies. The fire preventionist can also work with young people in group activities outside of school, such as in Junior Fire Marshal programs, Sparky fire departments, and the Scouts.

Publicity aids for fire prevention can take many forms: contests staged by the fire department; billboard advertising; distribution of signs and posters, book covers, and favors printed with fire prevention messages; innovative ideas such as talking trash cans or fire alarm boxes; or roving fire engines delivering messages through loudspeakers. The possibilities are endless.

One of the most effective means of reaching the public is through media publicity—the Internet, newspaper, radio, and television. The press should be encouraged to report fires accurately and to give specific information on contributing causes. This can have a tremendous impact on citizens of the community. Public service spot announcements on radio and television and television coverage of fire safety demonstrations and special events can all play an important part in making the community aware that it is everyone's duty to help in preventing fire.

■ ■

Review Questions

1. Public fire safety education activities include
   a. fire prevention
   b. fire reactions
   c. burn awareness
   d. none of the above
   e. all of the above
2. Wildfire mitigation education is the responsibility of
   a. local fire departments
   b. state organizations
   c. national forestry organizations
   d. state and national forestry organizations

3. Public fire education planning involves a _____-step process.
   a. 3
   b. 4
   c. 5
   d. 6
   e. 7
4. Which of the following is NOT considered home inspection duties?
   a. exit signs
   b. smoke alarms
   c. public relations
   d. potential fire hazards

5. Fire safety clinics have been valuable for which of the following audiences?
   a. nursing homes
   b. hospitals
   c. apartment house operators
   d. both a and b
   e. all of the above
6. "Learn Not to Burn" is a program from
   a. NFPA
   b. IAFC
   c. IAFF
   d. IBC
   e. USFA
7. Through the years there have been better chances of getting a fire prevention message to
   a. preschoolers
   b. youth
   c. teens
   d. college students
   e. seniors
8. Schools may not accept fire safety material if it includes
   a. health and wellness information
   b. crime prevention
   c. advertising
   d. toys or games
9. SAFE KIDS campaigns may include all the following except
   a. bike safety
   b. child vehicle occupant protection
   c. exercise program
   d. injury prevention
   e. fire safety
10. The wildland fire prevention teams used in Texas are designed for
    a. owner education
    b. businesses
    c. schools
    d. fire departments
11. Fire Prevention Day is always _____, which commemorates the Great Chicago Fire of 1871.
    a. September 7
    b. October 9
    c. October 24
    d. November 15
    e. December 7

■ ■

Answers

1. e
2. d
3. c
4. a
5. e
6. a

7. b
8. c
9. c
10. a
11. b

■ ■

Notes

1. U.S. National Commission on Fire Prevention and Control Report, *America Burning* (Washington, D.C.: U.S. Government Printing Office, 1973), p. 105.
2. Cathy Prudhomme, "Wild Wild West," *Fire Prevention—Fire Engineers Journal,* Institution of Fire Engineers (Leicester, U.K. 2003).
3. Federal Emergency Management Agency. U.S. Fire Administration, *Public Fire Education Planning, a Five Step Process,* 2001.
4. Powell Evans, comp., *Official Record of the First American National Fire Prevention Convention* (Philadelphia: Merchant and Evans Co., 1914), p. 150.
5. Ibid., p. 384.
6. Ibid., p. 150.
7. Percy Bugbee, *Men Against Fire: The Story of the National Fire Protection Association, 1896–1971* (Boston:

National Fire Protection Association, 1971), p. 20.

8. Personal communication to the author from Cheryl Edwards. Lakeland, Florida, Fire Department, 2003.

9. State Firemen's and Fire Marshals' Association of Texas, *Info Fire* (Austin, Tex., 1998), p. 1.

10. Personal communication to the author from Robert Allan, former North Dakota State Fire Marshal.

11. Bugbee, p. 39.

12. Philip Schaenman, Hollis Stambaugh, Christina Rossomando, Charles Jennings, and Carolyn Perroni, *Proving Public Fire Education Works* (Arlington, Va.: TriData Corp., 1990), pp. 113–115.

13. Ann Kulenkamp, Barbara Lundquist, and Philip Schaenman, *Reaching the Hard-to-Reach—Techniques from Fire Prevention Programs and Other Disciplines* (Arlington, Va.: TriData Corp., 1994).

14. Steve Bethke, "Public Perceptions in the Marketing Mix," unpublished paper, 1993.

15. Wingspread IV Conference Report, *The Fire and Emergency Services in the United States* (International Association of Fire Chiefs Foundation, Dothan, Alabama, Fire Department, 1996), p. 9.

Enforcing Fire Safety Compliance

4 CHAPTER

Foremost among the responsibilities of the fire prevention bureau is the enforcement of the municipal fire code. Some of the approaches that have proven to be effective in developing fire safety codes and regulations are examined in this chapter. Inspection is an intrinsic part of most of these approaches, but because the subject covers so much ground and deserves special attention, it is treated separately. Engineering principles have a major role in code development.

◆ PUBLICITY FOR FIRE CODES

Publicity regarding fire safety and building codes and the enforcement of such codes has not been used as often as it could have been to educate the public in fire safety. As one example of effective use, the fire prevention bureau should seek press coverage in connection with hearings for enactment of fire prevention codes. Although this may bring out individuals who are opponents of the legislation, publicity gives citizens a clearer idea of fire code coverage. Media publicity on code enforcement procedures is an excellent way to reach the public. Press publicity of code enactment is too often limited to occasions when a requirement is imposed to abate a specific hazard that has come to light as the result of a tragedy.

Signs placed in public buildings indicating that certain activities are violations of the fire safety code such as those indicating "No Smoking" or giving occupancy capacities such as "Maximum Capacity, 1,000—City Fire Prevention Code" remind citizens of fire safety. While potential dangers from careless smoking are obvious, maximum capacities may not be understood by many.

A year after the tragic fire at The Station nightclub in West Warwick, Rhode Island, which took 100 lives, only the state of Rhode Island has made sweeping fire safety rule changes for such occupancies. Several other states and cities have increased controls over indoor pyrotechnics. This again emphasizes the point that code changes aimed at strengthening requirements in a given occupancy seldom happen beyond the boundaries of the state in which the tragic fire occurs.[1]

◆ PLAN-REVIEW PROGRAM FOR FIRE CODE ENFORCEMENT

A number of municipal fire departments have programs for the review of plans and specifications for buildings to be constructed within their jurisdiction. Qualified individuals are assigned to check plans and specifications for conformance to fire and building code requirements. Usually, these reviews are coordinated with those carried out by the local building department. Plans are received originally by the building department and are then routed through the fire department as a matter of procedure.

A plan-review program can best be established through a municipal directive or state law stating that all plans for construction be reviewed by the fire department's fire prevention bureau. Florida, for example, has such a law.

The scope of plan reviews carried out under these arrangements is not always as broad as it might be. In a number of jurisdictions, the review is confined to checking the locations of fire hydrants, accessibility for apparatus, and of standpipe connections. Major items of concern in fire protection, such as exitways, interior finishes, basic construction, and sprinkler layouts, often are not checked by the fire department.

Personnel need professional training to carry out these responsibilities. They cannot be expected to participate in detailed reviews of plans and specifications and in conferences with architects, engineers, and builders without having an adequate background. Fire service personnel who have been assigned to plan-review duties without proper training and background have caused a loss of prestige professionally for the departments they represent.

FIRE PROTECTION ENGINEERS FOR PLAN REVIEWS

Ideally, a fire department will have a fire protection engineer available to carry out plan reviews, reviews of specifications, and consultations with architects, engineers, and builders. The desirability of having professional personnel who can speak with authority in the engineering field assigned to plan-review duties cannot be overemphasized. Smaller municipalities may find the addition of a fire protection engineer to the payroll prohibitive in cost. Over 100 fire protection engineers are assigned to fire departments in the United States. The fire protection engineer is usually in a civilian, nonuniformed force and in a staff rather than a line position.

In addition to having duties connected with review of plans and specifications, fire protection engineers are frequently concerned with long-range departmental planning relating to station locations, response districts, etc. They may also have considerable impact in the development of fire prevention codes and regulations. In view of their assignment in staff positions, they are usually not responsible for directing inspectional programs and other line functions, but are called upon to assist with highly technical code issues such as posed by performance-based codes. These codes have increased the demand for fire protection engineers.

In smaller municipalities, where the possibility of employing a fire protection engineer may be remote, the service might be more logically provided at the state level. The state fire marshal's office might have a fire protection engineer on staff who could be made available to assist with plan reviews in jurisdictions that find it difficult to provide an engineer of their own. This arrangement has an advantage from the standpoint of workload and is definitely a procedure by which best advantage can be taken of available talent. Another advantage to having the fire protection engineer at the state level conduct the plan review is that the review would encompass consideration of both state and local fire safety codes, as well as consideration of the sections of the building code that pertain to fire protection. Some small jurisdictions contract with nearby cities or with consulting firms to provide this service on an as needed basis.

Plan-review programs should, if at all possible, provide for review of preliminary plans. Some jurisdictions do not permit personnel to take the time to review plans in the preliminary stage; however, review at that point is usually rewarding. The entire direction of a project from a fire protection standpoint can often be fairly easily changed at the time of a preliminary review; however, a change at a later stage is usually considerably more difficult and many times more costly. Public relations are also improved by an early review of plans, specifically in the preliminary stage. Fire departments that do not conduct reviews of preliminary plans undoubtedly end up spending a greater length of time reviewing final plans than would have been necessary had a preliminary review been conducted, and problems that might have been easily corrected if detected in a preliminary review can develop into major problems at the time of a final review.

Most architects and engineers appreciate having a plan-review service available because it serves as an additional level of design review. Plans should be submitted in such form as to permit detailed review of the architectural and engineering documents.

Although consultations with architects during a plan review are necessary and desirable, the reviewer should have the opportunity to study the plans in detail alone, without interruption or distraction. Notes on the review should be kept on file, and narrative comments should be made to the architect. In some communities, the shortage of personnel has made it necessary merely to mark comments on plans and return them to the architect. This arrangement leaves a great deal to be desired because it permits no retrieval of notes and recommendations once the plans have been returned.

MICROFILMING OF PLANS

Some communities microfilm the plans submitted for review. This enables plan-review personnel to retrieve information months or years after the building has been constructed, which is especially desirable where alterations and additions may be made at a later date. Microfilming of plans also has the advantage of reducing storage space requirements. It is practical to retain the plans in their original form until the building has been constructed and then reduce them to microfilm. Computer-aided design permits retention on computer as well. An ability to retrieve plans is also desirable from a legal standpoint. Questions about details of approved plans may arise many years later, especially after a major fire in the building. Postconstructive shortfalls cannot be checked against review comments.

TRAINING OF FIRE SERVICE PERSONNEL
FOR PLAN-REVIEW FUNCTIONS

A community in which it is not feasible to hire a fire protection engineer, and for which the state or county does not provide such services, can have an effective plan-review program by preparing other personnel for such duties. In most fire departments, at least one person, possibly a firefighter with drafting experience, could be trained in reading construction drawings and in fire protection requirements. This at least provides some representation of the fire service in the plan-review process. It will be more difficult for such a person to gain acceptance by architects and engineers than for an engineer or fire protection graduate.

CODE BACKING FOR PLAN REVIEWS

In reviewing plans, the fire protection engineer or other person carrying out these responsibilities should have a designated code for reference to provide full backing for decisions. As previously mentioned, many building code requirements are applicable. If fire safety–related portions of building codes are to be enforced by a fire department review person, prior arrangement must be made with the building official to avoid conflict and misunderstanding.

PLAN-REVIEW CORRELATION WITH INSPECTION

Plan reviews are only as effective as the inspection follow-up to ensure full compliance with requirements noted by the reviewer. Requirements set down by letter from a plan reviewer or by red pencil marks on a set of plans may not always be carried out by either the architect or the builder. Unfortunately, omissions cannot always be readily detected in the field because the deficiency may be enclosed during construction and may not be revealed until a fire occurs. Inspections should be scheduled as necessary to coincide with the completion of various stages of construction so that the inspector can be at the site to detect any deviations from specifications.

Figure 4.1 ◆ This chalk building in Nebraska exemplifies the need for fire inspectors to be familiar with all types of construction.

COOPERATION IN PLAN REVIEWS

Cooperation among several agencies is absolutely essential for accomplishing the purposes of the plan review. These include the building department, the fire department, state-level agencies, and any other governmental agencies involved in the plan review. In a number of jurisdictions, all agencies having responsibility for plan review and code enforcement meet periodically to review plans and to discuss problems with architects, contractors, and others involved in the project. Some jurisdictions also include zoning, planning, traffic control, environmental services, and law enforcement representation.

A national campaign is under way to reduce by 60 percent the time it takes to move construction through regulatory processing from the initial zoning approval to the issuance of a certificate of occupancy. Over 50 national associations and federal agencies have joined with the National Conference of States on Building Codes and Standards (NCSBCS) to implement this reform. Many fire departments participate in "one-stop shops" where local review agencies function under one roof to facilitate the same goal of expedient processing of regulatory provisions.[2]

The insurance industry may also have an interest in reviewing plans for construction, especially in major industrial and institutional facilities. There is a possibility of a conflict between fire protection requirements of the insurance carrier and those of the fire department. Although it is presumed that the fire department's requirements must be met, on occasion arrangements might be worked out to encompass the requirements of the insurance industry as well through alternate procedures.

A requirement made by an insurance carrier reviewer is not necessarily out of order just because it is not found in the fire safety code. Of course, the fire department cannot place itself in a position of officially requiring adherence to provisions other than those for which it has a legal responsibility for enforcement.

◆ CONTROL OF SALES AND USE AS A MEANS OF FIRE CODE ENFORCEMENT

The control of sales and use is another means by which codes for fire abatement and prevention may be enforced. Most fire codes contain provisions that give the fire prevention bureau responsibility for controlling sales of explosives, fireworks, fire extinguishers, fire alarm systems, and other related devices, as well as general control of the handling of sales of gasoline and other hazardous materials. Maine and several other states require that all electrical devices sold in the state bear the seal of a nationally recognized electrical testing laboratory.

CONTROL OF GASOLINE AND OTHER HAZARDOUS COMMODITIES

Control of the sales and use of gasoline is found as a subject in practically all fire prevention codes. Included are required procedures relating to the handling, storage, and dispensing of gasoline, with specific requirements relating to the type of container into which gasoline may be pumped, how gasoline can be dispensed, the prohibition of smoking, and requirements that engines be turned off during dispensing.

Control of sales also includes sales, handling, and storage of other dangerous and hazardous commodities. In all cases in which the fire prevention bureau is involved, there is a possibility of ill feeling because no businessperson enjoys being told that a product is unsafe, not listed, or otherwise unqualified to be sold within the city.

FIGURE 4.2 ◆ Inspection of flammable liquid storage facilities requires specialized training.

Specific sales methods are also subject to control. An example is self-service dispensing of gasoline. This method of dispensing gasoline has been given a great deal of study by public fire service personnel. Self-service dispensing is very popular. It appeals to the customer because it is quick, and the economic factor is a valid concern of the service station operator.

Control of sales and use of hazardous materials are delegated to the fire department in many jurisdictions. Most fire codes specify which hazardous materials present a fire or explosion danger and include their control as a fire department function. Some jurisdictions assign responsibility for materials that present solely a toxicity hazard to some other agency.

Federal "right-to-know" provisions mandate that fire departments be notified of hazardous materials in their service area. Although the requirement is primarily to facilitate fire fighting, the data can be helpful to the fire prevention bureau as well.

Some jurisdictions require hazardous materials storage locations to be identified to assist firefighters. Generally, the fire prevention bureau is responsible for enforcement of these mandates.

CONTROL OF FIRE ALARM SYSTEMS

Control of the sales of fire alarm systems, including smoke alarms, is a public safety function. A fire alarm system is sold to the public for fire safety purposes and must be properly designed to function in the event of an emergency. From a public safety standpoint, the average person may have no way of knowing whether a fire alarm system, especially one designed to detect fires, is functioning properly until there is an emergency. In the case of practically every other type of household equipment, including toasters, TVs, and DVD players, the buyer can try the product at home to see whether it will perform satisfactorily. Some communities require that all fire alarm systems in the community be approved by the fire department. Most jurisdictions require the installation of smoke alarms in residences; some require installation only in new homes; others include installation in both new and existing residences.

CONTROL OF FIRE EXTINGUISHERS

Fire extinguishers are also regulated at the point of sale in some jurisdictions. Some substandard fire extinguishers are on the market. They have extremely limited value in the control of fires; however, they are frequently advertised as having superior fire suppression capabilities. Regulations enforced in some areas are designed to keep inferior extinguishers off the market.

CONTROL OF EXPLOSIVES

The control of handling, storage, and sales of explosives has long been covered in fire prevention codes in hundreds of municipalities, as well as in fire codes of counties and states. Explosives storage is considered in zoning regulations as well. The control of the sales of explosives may include requirements for identifying people who purchase explosives, as well as listing the names and addresses of purchasers. Control in this field is also a function of the federal government through the explosives regulations enforced by the Bureau of Alcohol, Tobacco and Firearms.

At the local and state level, explosives control is often considered a responsibility of fire prevention code enforcement agencies. Although explosives are actuated by fire or percussion, the logical avenue for control appears to be within the fire prevention field. For this reason, the fire marshal's office or fire prevention bureau has usually been designated as the agency for enforcing all regulations relating to the control of explosives.

Security Measures. Problems in the field were relatively simple until the wave of protest movements in the 1960s brought about the widespread use of explosives by militant individuals and groups. Regulations may control the sale of explosives to ensure that they are limited to bona fide purchasers and thus to preclude acquisition of explosives by militants and others with destructive purposes. Theft then becomes a major problem, and secure storage of explosives must be ensured. Regulations for storage magazines usually specify theft-resistant as well as fire-resistant construction and security locking devices. In some instances, security fencing may also be required. Terrorist activities have intensified the need for stringent security measures.

Deactivation Measures. The control of explosives may also require the provision of personnel to deactivate or otherwise render safe clandestine devices. The use of bombs and timed devices incorporating explosives has increased considerably, and control measures must be taken by the community to ensure that personnel are available to render devices harmless.

Several municipal fire departments and state fire marshals' offices have responded to this need by developing bomb squads for the purpose of handling explosive devices. State fire marshals in several states operate such squads.

Police Jurisdiction. In some communities the explosives control responsibility is under the jurisdiction of police agencies because an intentional explosion is a criminal act that is under police jurisdiction for investigation. There are some excellent police bomb squads.

Fire Department Jurisdiction. The assignment of responsibilities for explosives control within the fire service has the advantage of a close tie-in between inspection of storage locations and possible theft and misuse. In the event that responsibility for full control is assigned to the fire organization, personnel must be fully conversant in investigation techniques and have a knowledge of explosives.

Control of Fireworks. Another field of responsibility normally assigned to the fire service, but which often involves police agencies as well, is that of fireworks control. This control responsibility probably has as many varieties of approach as any subject in the fire prevention field. There are many differences in requirements relating to fireworks from state to state and even within counties in a given state.

A number of states do not permit the sale or possession of fireworks except for public displays conducted with the permission of the state fire marshal's office or local governmental agency. In other states certain types of fireworks may legally be sold over the counter. Generally, the "closed" states have provisions for fireworks possession only for displays sanctioned by the enforcing agency. Permits for such displays are usually granted only after a bond has been posted or insurance coverage is in effect to cover accidents that may occur during the event. By issuing these permits, the state sanctions an activity that would otherwise be illegal; therefore, special precautions must be taken to ensure safety.

Generally, the issuance of fireworks permits is a function of the fire prevention agency. Sale and possession of illegal fireworks is done in a clandestine manner, which requires detailed investigation to abate. This enforcement activity may be carried out jointly by fire and police personnel.

Fireworks are a danger primarily to personal safety, though they may also present a hazard with respect to fire safety. As in the field of explosives, training is needed to carry out an effective fireworks control program.

Control through Limitation of Sales. Control is usually effected through limitations on sales or requiring permits for possession; however, a major problem always exists when one jurisdiction permitting sales adjoins a "closed" jurisdiction. The public can simply cross the state line or municipal boundary to purchase fireworks and then return home. Public sale of certain types of fireworks is permitted in unincorporated areas of a number of states, but major cities and towns in these states do not permit possession of fireworks without a permit.

Enforcement of local regulations is difficult where items may be purchased nearby. Public education programs have been used in an effort to control this problem. In some cities, fire apparatus has been placed at the city limits with signs affixed to warn the public of the illegality of fireworks within the city. Loudspeakers may also be used.

In several jurisdictions, unmarked cars have been used to trail purchasers from the point of sale to a point within the city in which possession is illegal. A stop is then made, and action is taken to seize the material and arrest the perpetrators.

Legal Ban on Fireworks. Fire service organizations have been in the forefront among agencies attempting to require a legal ban on the public sale or use of fireworks.

In fact, fire service organizations are probably primarily responsible for restrictive legislation in the majority of states.

A federal law enacted in 1954 prohibits the transportation of fireworks into any state for use outlawed by that state. Offenders are subject to a maximum fine of $1,000 or a year's imprisonment or both.

Another organization that has had a major interest in fireworks legislation is Prevent Blindness America. This society is especially interested in eye injuries, but its effective work has not been limited to reducing eye injuries. The organization has successfully worked in a number of states to promote prohibitive legislation. This society works in cooperation with the National Fire Protection Association to enhance public awareness of the dangers associated with use of fireworks.

Public Fireworks Displays. Although public fireworks displays with permits are allowed in practically all jurisdictions, it is incumbent on the fire marshal's office to monitor those events. Provisions should be made for closing down the display in the event of problems. In some jurisdictions, individuals who set off displays are required to be certified by the fire prevention agency.

Enforcement of Regulations. Enforcing explosives and fireworks regulations brings fire prevention bureau personnel into contact with members of the public with whom they usually have little contact. The explosives control field, for example, involves a considerable amount of contact with the construction industry. A close relationship must be maintained with these individuals to ensure safe operations.

Likewise, fireworks control, if effective, may create a number of adverse situations. The fire prevention inspector may be placed in the position of having to seize fireworks and file juvenile petitions against middle school or high school students. Probably most other contacts by the fire prevention official with people in this age group are of a positive nature. The confiscation of fireworks that have been legally purchased in another jurisdiction is usually not well received and definitely places the enforcement official in an adverse situation. Because of this, there may be a reluctance on the part of fire safety personnel to follow through with full enforcement of the regulations.

◆ STRUCTURAL CONTROL AS A MEANS OF FIRE AND BUILDING CODE ENFORCEMENT

Structural control of buildings—another means of enforcement—is effected through the previously mentioned plan-review function coupled with inspection of structures, before and on occupancy, to be sure that all fire safety requirements have been fulfilled. This inspection may be conducted in company with building officials. Structural control includes inspection of materials, appurtenances, and other factors relating to the building.

FIXED MANUFACTURING EQUIPMENT

Fixed equipment used in processing and manufacturing is included within the scope of structural inspections for fire code enforcement. Some of this equipment has design features that may lead to fires. These hazards must be identified in the inspection process. On occasion the equipment may increase in degree of danger as it is used. Again, it is the responsibility of the code enforcement person to see to it that such equipment is noted and that proper repairs are made. Unfortunately, it is often difficult for the fire inspector to obtain proficiency in fire safety characteristics of major industrial machinery.

INTERIOR FINISHES

Control of the structure relates also to interior finishes. This is an important aspect of fire protection that has often been overlooked during fire safety inspections. Proper evaluation requires analysis of materials used in a structure to determine that the materials will not readily burn or, if they are ignited, that rapid spread of fire will not ensue.

FIXED FIRE PROTECTION EQUIPMENT

Inspection of fixed fire protection equipment is another part of the job of control of structures. Tests and inspections at the time of occupancy are also quite important. As with other phases of fire protection inspection, personnel assigned to this responsibility should be familiar with testing procedures necessary to ensure proper operation.

In some fire departments, fire suppression personnel are called on to assist fire safety personnel with these inspections and tests. Special equipment may be necessary to properly test fire protection devices. Frequently, water supply tests must be conducted at night and on weekends because of the problems created by the flow of water over streets and other public areas. It is always prudent to have the owner's representative perform the test with the inspector as a witness.

REINSPECTION

Control of structures must be maintained by continuing to inspect buildings to note changes that take place from time to time. The addition of a partition or change in the swing of a door, as insignificant as it may seem, could make a major difference at the time of a fire. All such changes can be noted in a structural control inspection program. Reinspection of buildings should also include inspection and supervision of testing of fire protection equipment on a periodic basis.

TEMPORARY CONDITIONS

Inspection for the control of structures must also recognize temporary conditions that may retard chances for egress in the event of a fire. This refers to inspections to determine that exits are not being blocked by furniture, cabinets, or other appurtenances and to determine that exits are not being locked when the building is occupied. This type of inspection is similar to that carried out for control of occupancy.

ELECTRICAL EQUIPMENT

Another phase of the inspection procedure relating to the control of structures is that of checking electrical equipment and devices. Although fire safety inspectors do not normally carry out a full range of electrical inspection duties, some elements of electrical safety inspection are noted by the fire inspector on routine inspection. Frayed wiring, obvious overloads, circuits not identified, extension cords, arcing equipment, and inadequate fusing are conditions normally found within the scope of the fire inspector's responsibilities.

◆ CONTROL OF OCCUPANCY AS A MEANS OF FIRE CODE ENFORCEMENT

Control of occupancy is another enforcement measure. This type of control includes determining and posting occupancy capacities within structures, coupled with the periodic presence of fire prevention bureau personnel in the facility to ensure that an excessive number of people do not occupy it and that all exits are usable.

An additional example of control of occupancy is the assignment of fire department personnel to public events to see that aisles are not blocked and that overcrowding does not occur. These duties are most effectively performed by personnel in uniform. Control of occupancy may also encompass programs designed to make sure that company employees know how to operate fire protection equipment on the premises. This would include the training of employees of places of assembly in the use of fire extinguishers.

People creating a disturbance in audiences have made the job of occupancy control a more difficult one. On some occasions, overcrowding has been reduced through effective control by fire personnel. It is recognized that the involvement of fire service personnel with an operation of this type does not improve public relations. People who are barred from entering premises because of fire code capacity requirements are often angry and may on occasion take direct action against fire department personnel. Many people resent being told what to do in a place of assembly, especially if they have paid to enter the premises. Extreme tact must be used at all times in this extremely important duty.

The fire department's activities in the control of occupancy may include efforts to ensure adequate fire reaction. There is a close relation between fire safety inspection and measures that contribute to life safety through instilling positive fire reactions.

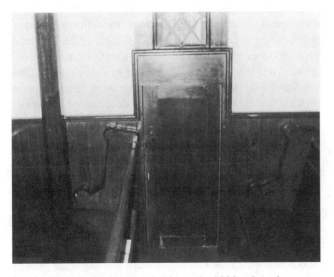

FIGURE 4.3 ◆ Builders of this early 1900s church provided extra exits between every other pew to offset the single main exit near the stove.

In compliance and abatement procedures for fire safety code enforcement, the ultimate goal is the improvement of the facility from a fire safety standpoint rather than imposition of fines or other punitive measures. There is little satisfaction in having fines levied or jail sentences imposed if in fact the facility remains unsafe. Court actions are time-consuming and should be considered as a last resort rather than as a primary means of enforcement in fire prevention.

Some fire prevention personnel have the authority to issue notices that serve as official notification of violations. In some cases inspectors may impose fines when they issue these notices. This procedure enables the fire department to immediately correct hazardous conditions and is somewhat analogous to the police officer issuing a traffic citation or ticket to an errant motorist. The logic is that a fire code violation is endangering others, and the culprit should be dealt with in an appropriate manner.

In addition to serving legal orders, fire departments may on occasion have reason to obtain injunctions or other sanctions to bring about prompt action to correct a serious fire hazard. Since procedures vary greatly from state to state, no attempt will be made to delineate specific legal processes. Fire department personnel having enforcement responsibilities should meet with the attorney for their jurisdiction to learn about specific avenues available for enforcement purposes. It is far better to obtain this information in advance of an incident rather than on the spur of the moment when life is in jeopardy in an occupancy.

Decisions handed down by several state supreme courts have a direct bearing on fire code enforcement. These are in addition to U.S. Supreme Court decisions cited in Chapter 5 with respect to proper identification of inspection personnel and permission to inspect.

In a number of such cases, courts have held that building and fire code inspection and enforcement are discretionary functions of other branches of government and that this entails no statutory duty of due care for the benefit of individual citizens or specific groups.[3] In one case, a state supreme court remarked that a failure in code enforcement represented no more legal liability on the city's part than there would be if a judge failed to make a decision or made a wrong one. Likewise, the city is generally not held liable for economic losses which may occur as a result of enforcement failures.[4]

In another case, the city was held not subject to recovery of damages in the deaths of and injuries to several motel guests who were trapped during a fire. Their escape was thwarted by improper stairway enclosures which were in violation of the code. Negligence on the part of the city inspectors in issuing the permit for renovation which included the improper enclosures was alleged.[5]

Courts have also taken the position that general negligence principles govern municipal liability where there is negligent performance of the building official's duties regardless of the "public" nature of these duties. These decisions would cover the performance of fire code officials as well. This means that fire prevention personnel cannot rely on any generally perceived immunity to protect them or their employers from successful litigation.

An appeal heard by the Kentucky Supreme Court, *Grogan v. Kentucky,* held that the city of South Gate and the Commonwealth of Kentucky were not liable for failure to enforce fire safety codes. The court held that government was free to enact laws without exposing the taxpayer to liability for failure to enforce these laws. The suit related to the Beverly Hills Supper Club fire that took 165 lives on May 28, 1977.[6]

On the other hand, courts have held that local government can be held liable for its actions when it issues a certificate of occupancy for a building with knowledge that the structure contains blatant safety and fire code violations. In one case, the court held that the town had a duty to refuse to issue a certificate of occupancy under such conditions and that such a rejection did not require the exercises of judgment or weighing of competing factors. The issuance of such a certificate creates a special relationship and duty to the owner or operation, the court held.[7]

The hiring of an incompetent inspector can likewise cause a municipality to be liable for damages. A village was subject to liability for its board's alleged negligence in having hired such an inspector. This inspector issued a building permit, construction began, and the inspector then revoked the permit for unspecified reasons. The court held that a municipality is just as liable as a private employer for harm caused by its negligence in hiring an incompetent servant.[8] Another decision held that the owner of a motel was responsible for building a structure in accordance with codes. A fire victim's representative had sued the motel for construction defects contributing to the fatality.[9]

◆ CONDEMNATION OF UNSAFE STRUCTURES

In addition to activities normally thought of as fire prevention measures, statutes in many states and ordinances in a number of cities provide fire and building enforcement officials with the power to condemn unsafe buildings. This power is granted to permit the fire or building official to order the demolition of structures that are found to be hazardous if the owner fails to take corrective action. Generally, the structures must be of combustible construction in order to impose the provisions of the statute or ordinance. It is presumed that a combustible building could endanger other properties if it caught fire.

Most condemnation statutes and ordinances also require that the building be situated in such a manner as to endanger other properties if ignited. In some jurisdictions, exposure to a telephone line or other utility line is sufficient to invoke the provisions of the statute.

These ordinances and statutes usually contain a requirement that the structure must also be unoccupied. There is also a requirement that the structure be open for trespass and therefore so situated that individuals can enter the building without difficulty. In some cases, arrangements may be made for boarding the building, thereby securing it in such a manner as to preclude the possibility of entry by trespassers, except by force.

The reasoning behind condemnation provisions is that structures open for trespass and of combustible construction that expose other properties contribute to the fire problem within a community. Such structures are open invitations to arsonists and are fire hazards. Many fires are set by juveniles in such locations.

Most condemnation ordinances contain requirements for the demolition or repair of structures in question, followed by the initiation of legal suit to recover costs. There may also be provisions for the recovery of court costs incurred.

Condemnation procedures have been aided by the willingness of fire departments to use structures for training purposes. Although this demolition method has worked well in a number of communities, it creates a possible conflict with air quality control requirements.

The condemnation of structures as a fire prevention measure is usually welcomed by other property owners in the neighborhood. Most buildings that are candidates for condemnation are unsightly and lower property values in the neighborhood. Some are frequented by drug users and other undesirable occupants. When the property is owned by an estate, it may be difficult to locate all the heirs to obtain service of the required notice. No work should begin in the demolition of the building until it has been definitely determined that the legal owner has been contacted.

◆ FIRE SAFETY CONSIDERATIONS IN SPECIAL OCCUPANCIES

Fire prevention and life safety measures in health care facilities and high-rise structures have all too often come about only when public concern is aroused in the aftermath of tragic fires. These areas are receiving more attention of late at all levels of government.

NURSING HOMES

An example of the evolution of fire prevention and protection requirements pertaining to a specific type of occupancy is that of nursing home fire safety. Because the potential victims are generally physically or mentally incapable of fending for themselves in the event of fire, the relative safety of nursing homes is based almost entirely on the effectiveness of code requirements.

The nursing home field is a developing one because of the increase in average life expectancy. There also seems to be a growing reluctance by families to keep older adult relatives in their homes when they can no longer actively participate in daily family life. Programs such as Medicaid and Medicare that provide financial assistance to older adults have likewise encouraged the development of nursing homes and other facilities for the extended care of the aged. The Medicaid program is primarily designed to assist those who are indigent, while the Medicare program provides assistance to those who have participated in the Social Security program through their work during more productive years. All of these factors have combined to create a great demand for nursing home care and other provisions for housing older adults.

As the industry expanded, little thought was given in some localities to the fire safety of the structures. In many cases, old, large dwellings that were no longer suitable for family occupancy were converted into nursing homes. In some jurisdictions, code officials equipped with adequate codes required the provision of automatic sprinkler protection and other safeguards in these facilities. In some cases strict requirements, such as sprinkler protection, enclosure of vertical openings, and other safeguards, caused nursing home operators to locate their facilities outside the city limits in an area where regulations were not so stringent.

A number of fires that resulted in multiple fatalities occurred in substandard structures in practically every section of the United States. In some states, these fires brought about modifications in fire prevention regulations, including requirements for automatic sprinkler protection and other fire safety features for all other nursing homes. Several jurisdictions found, only after a major nursing home fire, that there was no agency in the community or state that had responsibility for fire safety enforcement in such facilities.

The interest of the federal government in fire safety increased because of the previously mentioned programs for federal financial assistance to many older adult citizens. The United States Congress, as well as the Department of Health, Education, and Welfare (now Health and Human Services), recognized in the late 1960s that many of the patients receiving federal government subsidies were not being housed in safe environments. A number of states were not ensuring a maximum degree of safety for these people. There was also disagreement as to what were the best methods for providing adequate safety.

Fires resulting in multiple fatalities continued to occur, and the pressures to do something became greater and greater. Eventually, federal regulations went into effect in 1971 that required automatic sprinkler protection and other fire protection safeguards in many nursing homes. These regulations have had great impact in jurisdictions that had not previously required such protection. As could be expected, many of the same appeals were heard again, and it was necessary to discontinue funding in a few cases to enforce compliance.

Currently, the federal program is in full effect for skilled- and intermediate-care facilities. Enforcement of the regulations is federally funded and carried out by state personnel, normally through the state fire marshal's office or state health department.

BOARDING HOMES FOR OLDER ADULTS

Adequate safeguards for older adult residents of board-and-care homes, or boarding homes, as they are generally called, are far from universally provided. Many older adults are housed in old buildings of frame or ordinary construction, and very few are provided with automatic sprinkler protection or any other meaningful protective devices. A U.S. House of Representatives report on health care facilities made recommendations with respect to this problem, but progress in adequate safeguards will probably be limited until public interest is sufficiently aroused. The interest shown by Congress may serve to hasten action, but it is unfortunate that additional lives will probably be lost in these facilities before adequate safeguards are required in all jurisdictions. The problem is aggravated because control at the state level for facilities of this type is somewhat limited.

The U.S. Congress authorized a special study of health care facilities as a project of the Committee on Government Operations of the House of Representatives, which was carried out in 1972. The committee's report strongly recommended increased life safety for all health care facilities, including extended-care facilities and so-called boarding homes that house primarily older adults.[10] The committee recommended the continued use of federal financing as a lever to improve fire safety in health care facilities. During a recent 27-year period, almost 300 deaths in 64 board-and-care facility fires were documented by the National Fire Protection Association.[11]

HIGH-RISE STRUCTURES

A good example of how standards, procedures, and code requirements for fire prevention and fire protection evolve is the recent history of high-rise structures. Interest was focused in this direction as a result of multiple-fatality fires in high-rise structures that occurred in Atlanta, Georgia, and New Orleans, Louisiana, in the winter of 1972. The New Orleans structure was used for office occupancy with some public assembly use; the Atlanta structure was an apartment building for older adults. Both structures were quite modern, and construction was fire-resistive.

The fire in New Orleans gained national attention because of nationwide live television coverage. Through this exposure, a great many people became aware of a fire safety problem in high-rise structures. Before the advent of television this might have generated only local concern. The Atlanta fire called attention to problems in housing for older adults who are capable of living independently.

Both the Atlanta and New Orleans fires resulted in the immediate drafting of legislation, municipal ordinances, and regulations providing for fire protection in high-rise structures in a number of jurisdictions. In Maryland, for example, the General Assembly in 1974 enacted legislation requiring automatic sprinkler protection in all new high-rise buildings constructed in the state.

Louisiana, Massachusetts, and Nevada have enacted legislation requiring automatic sprinklers in existing as well as new high-rise buildings. Agricultural buildings are generally exempt from the provisions. Over 200 high-rise buildings have been sprinklered in Louisiana under these provisions.[12] Codes in some communities have been modified to require retroactive installation of sprinklers in high-rise buildings. Atlanta, Denver, Louisville, Philadelphia, and San Francisco are examples of this change.

In some cases tragic events have led to rapid enactment of fire safety codes. This has been part of the history of code development, as noted in Chapter 1. An example of this phenomena was the legislation enacted in the state of Nevada after the MGM Grand Hotel and Hilton Hotel fires. These tragic events led to the formation of that state's Governor's Commission on Fire Safety Codes. This commission studied fire safety and building codes in Nevada.

The commission met a number of times and reviewed pertinent codes. It also took testimony from fire safety experts and subsequently recommended enactment of a strong fire code that required automatic sprinkler protection retroactively in many Nevada buildings. Citizen pressure generated by these fires helped the commission get its findings rapidly adopted, resulting in major improvements in fire safety in publicly occupied buildings.

A similar chain of events occurred in the province of Ontario after a fire took six lives in the Inn on the Park in suburban Toronto in May 1981. The provincial government's exhaustive study of the fire led to a number of code changes.

Florida enacted a law in 1993 requiring full automatic sprinkler protection in all new buildings three or more stories in height. One- and two-family dwellings are not included in this requirement.

RESIDENTIAL SPRINKLERS

Another example of the code development process is that relating to residential automatic sprinklers. The development of mandatory requirements for residential sprinklers has taken a long and tortuous route.

While the value of automatic sprinklers has been recognized for over 100 years in industrial, commercial, and institutional occupancies, little credence has been given to the application of these principles to residential occupancies. The one exception was the use of automatic sprinklers in hotels and large apartment houses.

In the late 1970s the city of San Clemente, California, pioneered requiring automatic sprinklers in residences located beyond usual response limitations for fire apparatus. The city declared that residences built more than five miles from the nearest fire station had to be equipped with residential automatic sprinklers. The organizations developing the standards and the sprinkler manufacturers had to adjust to this new demand. The sprinkler industry began producing suitable equipment on a limited basis while a standard was developed and published.

The concept that "a man's home is his castle" has made this type of regulation difficult to implement. In addition, many home builders' groups have strongly opposed any measure that would bring about an increase in the construction cost of new dwellings. They feel that even a minor increase in cost will reduce the number of people who are able to buy new homes. Home builders point out that manufactured homes are not required to provide sprinklers, there by giving them a competitive advantage.

A number of jurisdictions have adopted regulations similar to those in San Clemente. In fact, San Clemente subsequently changed its regulation to require sprinklers in all new residences, a feature that is encompassed in most of the other residential sprinkler ordinances.

In 1985 Scottsdale, Arizona, became the largest city in the United States (with a population of over 148,000) to require residential sprinklers in all new residences. The city passed a requirement in 1974 requiring sprinklers in all new mercantile, industrial, and commercial structures with more than 7,500 square feet on the first floor or with two or more stories in commercial buildings or three or more stories in hotels or apartment complexes. This requirement has been so successful that efforts to convince elected officials as to the need and desirability of residential sprinkler requirements were made much easier. January 2001 marked the fifteenth year the city of Scottsdale, had required fire sprinkler systems in all new construction, both residential and commercial. As of the date the fast-growing city had over 39,000 single-family homes so equipped in addition to 19,000 multifamily living units. With a population of over 223,000, over 53 percent are protected with such systems.

The average fire loss for nonsprinklered structures during that period was $45,000 while sprinklered structures experienced a $3,500 average loss. More importantly, 13 lives were saved by sprinkler activation during the first 15 years the requirement was in effect. Installation costs for residences average between 55 and 75 cents per square foot.[13]

A number of cities in California are requiring automatic sprinklers in all new residences. A 2003 study in California indicated that 151 of 206 responding jurisdictions required fire sprinklers in all new residential occupancies including one- and two-family dwellings.[14] Such a requirement has been imposed in Montgomery and Prince George's county's Maryland, and many other jurisdictions.

Cobb County, Georgia, has also been a leader in residential sprinkler installation. Cobb County uses trade-offs rather than a specific ordinance as the method for securing sprinkler installations. The building official, with fire department support, has granted a number of trade-offs in multifamily residential construction requirements that result in such buildings being constructed less expensively with sprinklers than without. For example, two-hour fire rating separation requirements are reduced to one hour when the structure is sprinklered, resulting in a cost saving. Practically all contractors building multifamily residences in Cobb County are using this alternate means of construction, thereby effecting a savings in construction cost. Of course, the life expectancy of the building and its occupants from a fire safety standpoint is much greater, and the apartment units are more marketable.

Canadian cities are also requiring sprinkler systems in new construction, including residences. Westmount, Quebec, imposed this requirement in 1989, as did Quebec City. Vancouver, British Columbia, later became the first major North American city to require sprinklers in all new buildings.

An example of code development was the successful effort by the Florida Fire Chiefs Association and other organizations in that state to secure legislation that retroactively requires automatic sprinklers in existing hotels, motels, and time-share occupancies. This legislation was promoted by the fire chiefs with the support of the hotel and motel associations, League of Cities, and other interested groups. It came about without a multifatality fire in Florida as an impetus; however, the sponsoring agencies recognized the value of the tourism industry to the state and the devastating effect that such a fire could have on business. Previous attempts by the sprinkler industry to get this type of legislation enacted had failed, partly because they were looked upon as self-serving. A concentrated effort, plus a reasonable time period for compliance, brought about enactment in 1983.

Another consideration in the development of codes is that of evaluating new conditions, innovations, occupancies, and processes. As an example, the advent of the "hotel-hospital" brings a challenge from a code application standpoint. These facilities are designed for individuals who no longer require close hospital care but have recovered to the extent that they can exist in a homier, less expensive setting near the hospital. The latter feature is considered important because of the possibility of a relapse. The code developer and enforcer is concerned with appropriate fire safety measures for this new type of occupancy. Is it considered "institutional," or is it considered "residential"?

Another example is that of the development of the semiconductor industry. When faced with hazards of these occupancies, which were first developed in California, the western fire service was quick to develop a section of the Uniform Fire Code relating specifically to the semiconductor field. One problem in such a situation is that little empirical data are available to substantiate code requirements.

◆ SUMMARY

Adequate fire and building codes are not sufficient to provide fire safety. Enforcement is needed. There are many aspects of enforcement. Plan review by competent people is essential. In addition, fire code enforcement can be accomplished through control of sales and use, especially items such as explosives, gasoline, and fireworks. Structural control is also a means of fire and building code enforcement. This includes fixed manufacturing equipment, interior finishes, and fixed fire protection equipment. Communities have the ability to establish the legal right to impose fire safety and code requirements. Often it takes a disastrous fire to impact fire code enforcement. On rare occasions, some communities are proactive such as those that have adopted comprehensive sprinkler ordinances to protect their communities.

Review Questions

1. The Station nightclub fire in West Warwick, Rhode Island, was caused by
   a. careless smoking
   b. arson
   c. indoor pyrotechnics
   d. grease fire
   e. electrical short
2. The best person to conduct plan reviews for a fire department is
   a. fire protection engineer
   b. senior fire inspector
   c. architect
   d. building official
   e. battalion chief
3. Which department, in most cities, needs to cooperate closely on plan review issues with the fire department?
   a. police
   b. planning
   c. zoning
   d. building
   e. finance

4. Which is not typically an item controlled by sales for fire safety?
   a. explosives
   b. fireworks
   c. fire alarms
   d. fire extinguishers
   e. cigarettes and cigars
5. States that ban fireworks, except for public displays, are said to be
   a. open
   b. closed
   c. liberal
   d. conservative
   e. safety-strict
6. Inspections of structures before and on occupancy are considered
   a. structural control
   b. educational control
   c. plan review
   d. fire safety maintenance
7. Court actions for compliance should be considered as
   a. always necessary
   b. the responsibility of the fire chief
   c. a partnership with the police
   d. a last resort
8. The relative safety of nursing homes is based almost entirely on the effectiveness of code requirements because
   a. they are difficult to inspect
   b. the owners seldom cooperate
   c. the occupants cannot fend for themselves
   d. the staff is not trainable
   e. most fire department response is inadequate

■■■

Answers

1. c
2. a
3. d
4. e
5. b
6. a
7. d
8. c

■■■

Notes

1. "One Year after Deadly Club Fire, Few Changes Seen," *Tallahassee Democrat* (February 15, 2004).
2. NCSBCS news release, February 26, 1998.
3. *Trianon Park Condominium Association v. Hialeah* (1985, Fla.).
4. *Island Shores Estates Condominium Association v. Concord* (1992, N.H.).
5. *Hoffert v. Owatonna Inn Towne Motel, Inc.* (1972, Minn.).
6. *Grogan v. Kentucky* (Kentucky Supreme Court, 1979).
7. *Garrett v. Holiday Inns, Inc.* (1983, N.Y.).
8. *Lockwood v. Buchnan* (1959, N.Y.).
9. *Northern Lights Motel Inc. v. Sweany,* Opinion No. 1386, Alaska Supreme Court, 1977.
10. U.S. Congress, House, Sixteenth Report by the Committee on Government Operations, *Saving Lives in Nursing Homes* (Washington, D.C.: U.S. Government Printing Office, 1972).
11. Ed Comeau, "Board and Care Fires," *NFPA Journal* (September/October 1998), p. 33.
12. Communication to the author from V. J. Bella, Louisiana State Fire Marshal, 2004.
13. "15 years of Built-in Automatic Fire Sprinklers," unpublished paper, Scottsdale, Arizona, January 2001.
14. Stephen D. Hart, *Executive Summary Report on the 2003 California Fire Sprinkler Ordinance Survey* (Sacramento, Calif., 2003).

Fire Safety Inspection Procedures

CHAPTER 5

Inspection is a key function in the enforcement of fire laws and regulations. Duties of the fire prevention inspector call for knowledge and competency that can be acquired only through proper training and education in code requirements and inspection procedures. Other requirements are the ability to exercise good judgment, keen observation, and skill in dealing with people. As a representative of the fire department, the fire prevention inspector has the opportunity to build good public relations and to educate the public about the need to observe the rules for preventing fires.

This chapter outlines the basic principles that apply to the actual inspection procedures, whether conducted for control of structures, for control of occupancy, or for a combination of purposes. For more detailed information regarding inspections, excellent sources are publications of the National Fire Protection Association and the International Fire Service Training Association.

◆ PREPARATION FOR INSPECTION

Preparation should include instilling a positive attitude on the part of the inspector. The inspector must know why the inspection is being made and what to look for during the inspection. This is an important phase of preparation, one that, unfortunately, is often overlooked in training and in planning for inspection programs. Another requirement is the acquisition of necessary equipment, such as flashlights, cameras, notebooks, data loggers, and suitable clothing. In some departments the inspector is also required to carry manuals, code books, and other publications that may have a bearing on the work. Laptop computers are useful for this purpose. It certainly is a good idea to have these materials in the car for reference if needed.

KNOWLEDGE OF CODES

One of the prime requisites for competence in the inspector's duties is a thorough knowledge of the local fire safety code, as well as other relevant municipal or state codes that might apply. The inspector may not necessarily be fully conversant with each point in the fire prevention code but should be familiar enough with the provisions and requirements to make general observations and recommendations during the inspection tour, with subsequent detailed study where necessary. This might be done at the site by referring to code material carried in the car. In some instances, special problems may require more detailed analysis or consultation with supervisors after the inspector has returned to the office.

TIME REQUIREMENTS

Under no conditions should an inspection be undertaken when the inspector is pressed for time and is thus unable to make a thorough inspection. Fire safety inspections made under time constraints would in most cases better serve the cause of fire prevention had they not been made at all. An official inspection, however haphazard, may lull the owner or property manager into a false sense of security and subject the inspector and fire department to legal action.

Although several major insurance organizations have experimented with time quotas for conducting fire prevention inspections, this is a risky practice to follow. The complexities of building inspections and the varied degrees of understanding on the part of the building representative accompanying the inspector, as well as a number of other factors, combine to make it most difficult to stipulate a fixed amount of time for a fire inspection. An estimate of the average time needed for a given occupancy might be established, but under no conditions should an effort be made to limit the inspection to that figure. Thoroughness cannot be overemphasized. It is necessary to cover adequately all areas of the structure even though entry to certain spaces may be difficult. In many buildings, there may not be a master key for all closed areas, and certain spaces may be assigned to personnel who carry individual keys. This necessitates a search for the individual who has the specific key, which slows down the entire procedure. It is essential to take the time to obtain entry to all spaces.

INSPECTION FREQUENCY

Inspection frequencies vary considerably from one jurisdiction to another. Availability of personnel is a major factor in determining the amount of time between inspections. Another factor is the perceived degree of hazard in the occupancy. For example, institutions are usually inspected more frequently than storage warehouses. The potential for finding deficiencies may be a factor in scheduling inspections even within a given class of occupancy. Those facilities in which violations are usually found may be scheduled for more frequent visits than those where fire safety problems are seldom found.

North Carolina has mandated that inspections be conducted on a scheduled basis. Subsequent to the Hamlet food processing plant fire on September 3, 1991, the state's Building Code Council added a mandatory fire safety inspection schedule requirement to the Fire Prevention Code. Local governments are responsible for conducting these inspections of occupancies.

Hazardous, institutional, high rise, assembly (except churches and synagogues), child care, and common areas of residential occupancies must be inspected at least annually. One- and two-family dwellings are exempted. Industrial and educational occupancies except public schools must be inspected at least once every two years. Public schools are required to have at least two fire inspections per year. Business mercantile, and storage occupancies, churches, and synagogues require inspections at least every three years. Local jurisdictions may require more frequent inspections.[1]

Another measure instituted subsequent to the Hamlet fire was the State Department of Labor's adoption of the National Fire Protection Association's (NFPA)

Life Safety Code as part of its regulations. Most places of employment are covered by these provisions.

INSPECTION PRELIMINARIES

Before leaving the office or fire station, the inspector should review reports of any past inspections at the facility. These reports will be of material value in conducting an inspection and must be considered in evaluating the facility. A property owner is certainly justified in being alienated by a fire inspector who fails to recognize past efforts to correct deficiencies. Most property owners are proud of efforts they have made to correct deficiencies, and their work should be recognized. By reviewing past reports, the inspector can also get a better idea of the construction and general layout of the building, and the time taken to review the file will be rewarded by producing a more expedient inspection.

If the previous inspection was made by another inspector, it may be necessary to confer with that individual to ensure a clear understanding of all items listed in the inspection report. If the inspector going out on the job finds some previously reported items confusing, the property owner is most likely also having difficulty understanding what is meant. Furthermore, the last inspector may have missed some items.

MANNER OF DRESS

There is some interest in plainclothes dress for fire safety inspectors. On entering the premises, a person so dressed is not as apt to create a feeling of resentment on the part of the owner or operator as would someone in uniform. They are also not likely to be noticed by customers, a pleasing factor to the proprietor. Some form of identification is of course necessary. Appropriate identification cards or a blazer with an emblem or badge may be used. Many departments feel that a uniform is more effective in gaining compliance, however.

TAKING NOTES

The inspector should be prepared to carry a notebook, clipboard, recorder, or computer (personal data assistant) to make notes of all findings during the inspection. In any case, the notes taken should be orderly and easily understandable after a full day of inspecting many types of facilities. Specific locations of items needing correction should be designated in the notes, rather than general statements that might prove difficult to understand. There should never be an indication to the inspector's escort that notes are being kept in a careless manner. If an inspector uses the back of an envelope for notes, it gives the guide the impression that a thorough, accurate inspection is not being made. The inspector should be aware that the report will become a legal document in court.

◆ IDENTIFICATION AND PERMISSION TO INSPECT

The approach to and contact with the owner or operator of the property are important factors in establishing a favorable climate in which to conduct the inspection. On entering the structure, the inspector should immediately present identification to personnel in the building. If at all possible, the manager, superintendent, or other person in authority should be notified. Such identification should be made to ensure compliance with the guidelines of the U.S. Supreme Court decisions in cases pertaining to right of entry (*See v. City of Seattle* and *Camera v. City and County of San Francisco*). The decisions handed down in these cases make it imperative that the fire inspector present adequate identification to the proper personnel and request permission to make the inspection.

The cases referenced substantiate the property owner's right to refuse the inspector entrance to nonpublic areas except in situations involving imminent danger. The inspector must obtain a warrant to enter nonpublic areas when refused permission to enter.

On occasion a delay will be encountered because those in authority at the property may not be available at the moment the inspector arrives. The inspector should await the arrival of the person in authority and then present identification and request permission to conduct the fire prevention inspection.

Under no circumstances should the fire inspector proceed to make an inspection without proper identification and permission. It might be tempting to go ahead and make an inspection with the agreement of a janitor or other person not responsible for the facility. This can prove embarrassing, because it is generally considered an intrusion by management and may even result in management's taking punitive action against an unauthorized person for permitting the entry of the inspector.

When permission is granted to make the inspection, the customary procedure is for the inspector to request that a representative of management go along on the inspection tour. In most cases management readily agrees to this procedure.

Occasionally a response to an alarm by a fire department brings about an opportunity for observation of fire code violations. Response to a false alarm gave the Toledo, Ohio, Fire Department an insight when it found sprinkler system plumbing to be rusty and in need of replacement. The subsequent violation notice was appealed on the grounds of an allegation that the department's entry for the inspection was in violation of provisions of the U.S. and Ohio constitutions requiring search warrants. Because the entry was not for the purpose of finding evidence for a criminal prosecution, the appeal was denied.[2]

◆ THE INSPECTION TOUR

For many years it has been customary to start an inspection at the top of the structure. This should include a walk out on the roof to check roof structures with respect to fire safety. The inspector should check the elevator machinery room and should note whether the roof is used for occupancy. If it is, a check should be made to ascertain that proper egress exists.

The inspector and the escort (or escorts) should observe the condition and construction of the roof as well as make note of any fire protection problems that may exist at that level. This is also a good opportunity to view adjacent structures to evaluate any exposure hazards to the structure being inspected. At some point the inspector should also look around the outside of the building.

The inspection party should then move to the top floor of the building and make a thorough inspection of each section of that floor. Progress through the building should be systematic, and the same principles should be applied down through the entire structure, including the basement levels.

The inspector should be aware of any diversionary tactics that might be used to keep him or her out of certain areas. There may be an effort by the escort to steer clear of hazardous areas or those in which conditions may not be found up to standard.

Attempts to detour the inspector may not necessarily be made to avoid exposing hazardous conditions but may be based on a desire to keep secret certain proprietary processes, machinery, or research work. Industry often has a great deal at stake in the development of new processes or products. An unthinking fire inspector can cause a great deal of hardship to an industry by passing on such proprietary information. Escorts may need to obtain special permission for the fire inspector to go into certain spaces in which these processes are being carried out. This may mean a delay; however, it is important that the inspector wait for permission.

In making the tour through the building, the inspector should make a point to think about everything noted. One of the major weaknesses in fire inspections is that many inspectors look at hazards, unusual conditions, or exit deficiencies but somehow what they see does not register in their minds as a problem. Studies of major loss of life fires are replete with examples of this phenomenon. This is not necessarily a result of inadequate training in inspection procedures; understanding the significance of what is observed is a skill that takes practice to develop.

The building must be considered for possible development of a fire within it and from the standpoint of potential spread of fire to other structures. Careful consideration must also be given to anticipated occupant reaction in the event of a fire.

Conversation with escorts can cause the inspector to overlook important items. If the person accompanying the inspector is carrying on an extended conversation about sports activities or television programs, for example, the inspector can easily walk by a hazard without taking notice.

Although it is impossible to determine every factor that must be considered in examining a building, the salient points for inspection may be classified as follows: exposure hazards, potential fire hazards outside and inside of the building, potential fire causes, potential avenues for spread of fire, water supply, life safety features including means of egress, and alarm and extinguishing systems.

POSSIBLE EXPOSURES

The inspector should take full note of all appurtenances to the building, as well as exposures and hazardous locations nearby. Time spent in looking around outside to size up the proximity of other structures or combustibles that might, if involved in a fire, endanger the building in question can be most helpful to the inspector in assessing the overall fire safety of the facility. The inspector should also take into account the construction of adjoining buildings, the relationship of windows and other openings between the two structures, the existence of protection such as fire sprinkler systems within the exposure, and the height and area of the structure being inspected. There is a possibility that corrective action in the form of outside fixed protection, wire-glass windows, or other safety features might be recommended in the fire prevention report as a result of this review of conditions outside the building.

POTENTIAL FIRE HAZARDS

The fire safety inspector should check for the existence of combustibles that, if ignited, could cause the spread of fire. No blanket procedure can be recommended in this connection because of variables that must be considered. Hazards that present a danger of fire include flammable and combustible liquids and gases, oxidizing agents, explosives, acids, materials subject to spontaneous heating, combustibles, and explosive dust, as well as ordinary combustible materials in which rapid fire spread might be anticipated. The last category includes foam rubber, plastic packing material, cotton batting, paper, and other finely divided material that might cause rapid fire spread. The inspector must realize that all of these hazards may exist as a necessary and integral part of the operation of given occupancies.

The fire inspector may expect to reduce the number of hazards in some locations; however, in others the answer may involve providing fire sprinkler protection, separating fire hazard areas to reduce fire spread, or other steps designed to reduce the possibility of a serious fire developing. But the inspector must be resigned to the fact that it is impossible for most occupancies to operate without having at least some potential hazards on the premises.

The fire inspector must be able to evaluate problems created by combustibles in light of the potential fire spread. Certainly, a clothing factory, for example, cannot be told to eliminate all combustible materials. However, the extent of storage of combustibles in any location within the factory is directly related to protection available, construction, and possibilities for ignition of materials. The trained inspector must use judgment in deciding on appropriate action in each inspection. Combustibles are essential to the operation of many industries and mercantile establishments.

POTENTIAL CAUSES OF FIRE

During the course of an inspection, the inspector should observe sources of ignition that could cause a fire and take steps to reduce this danger to a minimum. Again it must be realized that it is impossible for the fire safety inspector to effect safeguards that would completely eliminate possible fire causes.

Sources of ignition that are frequently responsible for fires include electrical arcs and sparks; lightning; static electricity; friction through grinding, polishing, cutting, and drilling operations; various types of chemical reactions and open flames; improper heating devices; smoking; use of torches; and many other causes. Sources of ignition are present in most occupancies.

Certainly, an inspector should exert every effort to reduce fire causes in a structure to an absolute minimum. Conditions can change very rapidly; in fact, it is entirely possible that "cause" conditions abated at one moment are back again almost as soon as the inspector has left the building. As an example, a fire inspector may admonish an individual for smoking in a hazardous area only to have the person light up again as soon as the inspector is out of sight. Fire prevention inspection and control should therefore emphasize the need to restrict ignition sources to actual operating requirements, provide safeguards in installation and handling, and clear surrounding areas of combustible materials in which fire could spread.

POTENTIAL SPREAD OF FIRE

During the inspection tour, it is quite important for the inspector to consider the ways in which fire might travel within the building. Structural factors that might contribute to the spread of fire are the location and condition of elevator shafts, stair towers, light wells, laundry and trash chutes, pipe chases, conduit openings, vertical and horizontal ducts, dumbwaiters, combustible ceiling tile, combustible interior finishes, windows that might permit vertical spread of fire outside the building, shafts, and any other unprotected openings. Access to adequate means of egress is a major consideration.

Major causes of horizontal fire spread include the absence of fire division walls and partitions in basic construction and lack of or improper fire doors in existing fire walls. Combined with these structural factors, the presence of the combustibles and hazardous materials, poor stock storage methods, and lax housekeeping practices contribute to the spread of fire. The inspector with a fire suppression background is in a much better position to accurately evaluate fire spread factors than is one who is not familiar with fire fighting procedures and fire spread characteristics.

To make a proper fire safety inspection of the aforementioned points requires a thorough knowledge of building construction as well as an understanding of the travel characteristics of fire. Many fire inspectors unfortunately do not consider possible means of fire spread within a structure. Their trip through the premises is primarily for the purpose of noting observable conditions that might cause a fire, such as the presence of combustibles, careless smoking, locked exits, and careless handling of flammable liquids. The mark of a trained, competent fire inspector is an ability to comprehend fire spread possibilities inherent in the design and construction of the building.

WATER SUPPLIES AND EXTINGUISHING SYSTEMS

In touring the building, the fire inspector should check on water supplies for fire suppression purposes and observe standpipe connections, sprinkler systems, and any other fixed extinguishing systems.

In checking sprinkler and standpipe systems, every effort should be made to determine that the systems are functioning properly. Although many inspectors do not consider this a part of the inspection function, it should be a mandatory requirement of every fire safety survey. The fire inspector should not open the valve or otherwise test the device but should request that the building operator or owner's representative do so. Liability is assumed in actually operating a valve, and the facility operator should assume this obligation. The operator has a responsibility to satisfy the fire department that all fire protection devices on the property are operable. One major

FIGURE 5.1 ◆ Fire departments need to be familiar with all elements of public water supplies.

fire insurance organization believes that if its inspectors check only to see that sprinkler systems are in service, well over 50 percent of their desired results have been accomplished.

◆ CORRECTING VIOLATIONS DURING INSPECTION

As many items as possible should be corrected while the inspector is on the premises. It does not make sense for an inspector to record items pertaining to blocked fire doors without getting corrective action while on the premises. It is hard to justify not doing so when violations can be easily and rapidly corrected by the person accompanying the inspector. Another example is that of permitting a sprinkler system valve to remain in a shut position rather than having it opened. However, there may be times when the system valve is shut for some valid reason, but this should be checked out and verified and provision made to return the system to operating condition as soon as possible. In the prepared report, the fire safety inspector should mention items that are corrected during the inspection as a reminder to management to avoid repetition.

◆ DISCUSSING FINDINGS WITH ESCORTS

In some cases it is desirable for the inspector to discuss specific problems as they are discovered on the tour. In other cases it would be best for the inspector to record findings and then discuss the locations at the completion of the inspection. A major factor for consideration is the position in the firm of the individual or individuals going around with the inspector. There is no reason to go into a detailed explanation of specific locations of violations with a person who is not responsible for compliance, such as an usher in a theater. However, if the safety director for an industrial plant were accompanying the inspector, it would be desirable to point out and discuss the violations during the tour.

◆ THE EXIT INTERVIEW

If someone other than the owner or property manager has accompanied the inspector on the inspection tour, arrangements should be made for an exit interview with a responsible management representative to review and discuss the findings of the inspection. The person in authority may not be immediately available for an interview, and the inspector may need to come back on another day. In any event, the exit interview is an important part of the inspection and should not be overlooked.

During the interview, the inspector should advise the management representative of all the problems noted during his or her visit. If it is necessary to return to fire headquarters for further technical information, arrangements should be made to return or to telephone to discuss these matters in person or by telephone.

It is unwise to impose oral requirements that are not later substantiated in the prepared report. The property owner may make changes based on orally imposed requirements only to find that the inspector, after returning to the fire station, decided they were not necessary. In one case an inspector imposed a requirement for converting windows adjacent to an outside fire escape at a nursing home to wired-glass windows.

When the final report on the inspection was submitted, no such requirement was imposed. However, in the several weeks intervening between the inspection and the submission of the report, the conscientious nursing home operator made the changes at considerable expense. Needless to say, the operator was quite disturbed.

The lesson here is that if on return to headquarters the inspector finds justification for making any substantial changes in recommendations after they have been reviewed in the exit interview with management, the changes should immediately be communicated to the person interviewed, by either telephone or personal visit.

Requirements imposed by the fire inspector should be so clear and concise that they adequately portray to an untrained person the true picture of what is needed for compliance. In explaining the conditions during the exit interview, the inspector should give the owner or property operator the opportunity to offer suggestions on corrective actions. Compliance with requirements will often be more readily attained if the owner or property manager comes up with the solutions to the problems.

Technically, it is sufficient for the inspector merely to state that certain things must be done because the code requires it under a given section. This does not give the property management much satisfaction or understanding of the real need to comply with the fire code for life safety and protection of the property. By explaining as clearly as possible the reasons behind provisions of the code and how they apply, the inspector will make code enforcement much more palatable to management. It may not be necessary to go into great detail on all points, but at least a summary explanation should be given. The inspector must refrain from conveying a threatening attitude about the necessity of fire code compliance. The inspector who settles for "I don't know why but the code says it has to be done" leaves the impression of not being well trained or not sold on the importance of the inspection.

The property owner may be required to spend a great deal of money to comply with fire inspection requirements. The fire inspector has a professional obligation to be sure that all requirements will remedy the situation and not call for unnecessary expenditures. One foolish or impractical recommendation can offset the justifiable aims of the inspection. Management that is not pleased with the inspection will take full advantage of any improper or impractical requirement in attempting to downgrade the entire report. The inspector should clearly identify items that are merely recommendations as opposed to those that are mandatory. This becomes a major issue when "grandfather" clauses negate the "mandatory" classification of obvious hazards.

The manner in which the exit interview is conducted has important public relations aspects for the cause of fire prevention. A knowledgeable and friendly presentation of the findings can go a long way toward making management receptive to compliance with the code.

◆ REPORT OF INSPECTION

A report of all inspections for transmittal to the property owner or occupant is necessary. In some cases the report is handwritten by the inspector or prepared on a laptop computer and left on the premises. This practice is not recommended for larger facilities because it forces the inspector to state in writing all of the problems on the spot with no opportunity to return to the office to look up detailed requirements or possible alternate solutions. In some jurisdictions a completed inspection form is left at the premises.

A report in letter form gives more authority to the recommendations and makes a better impression from the standpoint of public relations. This depends on having report preparation resources available in the fire department.

The inspector has good reason to return to the office to refer to documents for technical information, to give more thought to the requirements and requirements specified, or to confer with supervisory personnel on the more complex problems. One disadvantage, of course, is the greater time lag between the completion of the inspection and the time the property owner receives instructions for compliance with inspection requirements. There is always the possibility that a fire or panic situation might occur during this intervening period; however, the probability of this happening is somewhat remote, and it would seem worthwhile to take the time to prepare a report, with copies of all correspondence to be kept for reference for subsequent inspections or possible legal action.

It is extremely unwise to make a statement or to include in a report words to the effect that the building is free of hazards. There are many potential fire conditions that could bring about a serious fire that even the most competent fire inspector could not be aware of, because they are concealed behind walls, in enclosures, and in other locations where observation is impossible.

Some larger fire departments have fire prevention bureau personnel who are specialized in certain phases of fire prevention work—for example, specialists in flammable liquids and gases, electrical, and other fields. The fire inspector on general assignment may consult with a specialized inspector in preparing the report, or a review of the inspection report may be made by a supervisor.

◆ REINSPECTION AND PROCEDURES TO ENFORCE COMPLIANCE

Normal inspection procedures followed by most municipal fire departments include a reinspection to ascertain compliance with the fire inspection report. An approximate date for the reinspection should be indicated in the report.

Procedures to enforce compliance with the fire code require the submission of a report of violations and recommendations for corrective action, as described in the preceding pages. The report usually includes mention of a specific number of days within which the work must be completed. Usually, the dates for anticipated completion take into account the complexity of the job, availability of materials, and other related factors. Housekeeping violations, for example, would normally be given a very short time to correct, while the installation of a stair tower would be given an extended length of time. A reinspection is made at the end of that time to determine whether the work is underway or has been completed. In some cities, the reinspection is considered the beginning inspection for any legal action taken to enforce compliance. If corrections have not been made, the property owner is given a formal legal order to complete the work within a given number of days or face a possible fine or jail sentence. Normal procedure includes obtaining a warrant for the arrest of the individual who has not completed the corrective requirements within the time specified in the legal order.

◆ CLASSIFICATION OF HAZARDS

In the category of hazards, a breakdown between those referred to as *common* fire hazards and those referred to as *special* fire hazards is usually made. The common hazards are those found in practically every occupancy, such as those related to normal use of electricity and heating equipment. Smoking is included in this category. Special hazards are those that the fire inspector associates with conditions unique to a given occupancy.

♦ HAZARDS IN VARIOUS TYPES OF OCCUPANCIES

In reviewing occupancies subject to inspection, particular hazards and causes characteristic of each may be noted. Familiarity with these hazards and causes is helpful to the fire safety inspector in knowing what to look for in each type of occupancy.

INSTITUTIONS

Long considered the highest risk occupancy from a life safety standpoint, the institution may present a number of unusual situations. A hospital, for example, will contain flammable liquids and gases as well as moderate quantities of combustibles in storage areas. From a life safety standpoint, the institution presents a major evacuation problem in that a majority of the occupants are not usually able to respond unaided to a fire alarm. Personnel within the facility, aided at a later time by fire department personnel, will need to evacuate such a structure or move patients to a place of safety. Common hazards found in such facilities include mattresses, bed clothing, and other combustibles normally found in patients' rooms. The clothing worn by the patient may become a hazard if it is not made of a fire-retardant material. The possibility of a fire occurring as a result of careless smoking becomes greater in a hospital or nursing home because occupants may be immobile.

Other potential fire causes likewise become magnified under institutional conditions. The use of a small amount of flammable gas, necessary in certain medical procedures, becomes a major hazard around electrical appliances. The possible results of an explosion and fire in an institution may be much greater than in an industrial occupancy.

PLACES OF PUBLIC ASSEMBLY

Another classification of occupancy generally associated with life safety problems is the place of public assembly. Large concentrations of people and the possibility of panic make fire code inspection of this type of occupancy especially important. Conditions in a large public auditorium are quite different from those in an amusement park or a restaurant.

A small fire in the stage area of an auditorium that would have little chance of spreading could cause panic and large loss of life under conditions of improper egress facilities, improper lighting, or overcrowding. The fire inspector conducting a routine inspection may visit a place of public assembly during a time of very limited occupancy. It may be difficult to visualize conditions that would exist if the place were filled with people. Public assembly occupancies also include art galleries, grandstands, tents, libraries, exhibition buildings, funeral homes, churches, skating rinks, transportation terminals, dance halls, and theaters.

Large civic centers and arenas capable of seating thousands of people for special events present special problems. Hazards and causes vary considerably in this broad occupancy classification. However, the one overriding concern is for assuring a means of immediate egress in the event of fire or other emergency. The possibility of panic always exists because there is no way of predicting the reaction on the part of an audience. This possibility again compels the fire inspector to reduce fire hazards and causes to the minimum.

In one major incident a small amount of liquefied petroleum gas, estimated at approximately three pounds, was responsible for taking 75 lives in the Indianapolis Coliseum explosion in 1963. In this case, the gas cylinder was located in a storage room adjacent to a refreshment stand inside the structure. Gas apparently escaped and was ignited as a result of a heater for popcorn that was located nearby. The explosion in the confined storage area ripped out heavy concrete and wreaked havoc in the building. The storage of liquefied petroleum gas could have been acceptable from a fire safety standpoint at some other location, but not in a public assembly occupancy. This structure was protected by a well-equipped fire department, and structural deficiencies did not contribute to the disaster.

In public assembly occupancies, fire safety inspectors have usually been successful in getting corrective action to eliminate major fire problems. Constantly recurring conditions that increase the likelihood of problems are overcrowding and locked exit doors. The tendency of some groups to attempt to enter facilities of this type without proper authorization may bring about severe overcrowding and may result in locking of exits to preclude illegal entry.

In restaurants, major fire hazards result from deep-fat frying and spattered grease. Protection systems are available for hoods and ducts; however, many older restaurants have not installed such protection. Lethargy in cleanup procedures can result in a hazardous accumulation of grease, which provides immediate fuel if ignited.

The fire in West Warwick, Rhode Island, that took 100 lives in the Station entertainment venue has again focused attention on public assembly occupancies. The February 20, 2003, fire has resulted in major code changes in Rhode Island. As in other tragic fires, the resulting code changes often occur only in the state in which the incident happened.

MERCANTILE OCCUPANCIES

Mercantile occupancies usually contain many fire hazards and causes. The trend toward large undivided areas for product merchandising has increased the possibility of major fire losses in retail stores. There is also the possibility of delay in notifying the fire department because a fire may not be readily detected in a large area.

Common hazards increase in number in such merchandising outlets, and the counters may contribute a great deal of fuel to any fire. Fortunately, a majority of large retail stores, generally those in excess of 15,000 square feet, are required to install fire sprinkler protection at the time of construction. In considering the overall potential loss from a fire in the mercantile establishment, the fire inspector should consider water damage. Because merchandise that has been damaged by water usually cannot be sold, the inspector might point out measures that could be taken to reduce the possibility of such damage. This kind of advice may not be required under the code; however, it can be quite helpful in reducing loss in the event of a fire. It may be feasible for the store operator to place merchandise in the storeroom on skids or pallets to reduce water damage potential.

Flammable liquids and gases may also be found in mercantile occupancies. Common causes such as defective heating equipment and smoking may be factors in fires in such occupancies. Retail stores may have special hazards associated with the occupancy. An example is a hardware store, where there is a good possibility of black powder being present. Another example is a store where charcoal briquettes are sold. This material is subject to spontaneous heating.

STORAGE OCCUPANCIES

Closely associated with the mercantile occupancy is the warehouse. If material that is stored poses a special hazard, such as large quantities of aerosols or combustible fibers, additional fire problems are created. Normally, warehouses and other storage occupancies are subject to in-and-out human occupancy. Therefore, there is not as much chance of a fire being started by human carelessness. However, this advantage may be offset by the possibility of delayed discovery because of limited time of occupancy.

Problems of water damage become even more acute in the warehouse, and serious consideration should be given to provisions that will preclude water damage if sprinklers or hose lines are operated. This may not prevent all water damage, but at least damage in areas away from the immediate fire area will be abated. Outdoor storage facilities such as lumber yards and coal yards serve the same purpose as warehouses. These facilities may be subject to fires resulting in large losses and may even be more readily ignited because of their accessibility.

OFFICE BUILDINGS

Office buildings constitute yet another type of occupancy. Common hazards are usually fairly easy to find in these structures. Common causes are also quite prevalent. A major problem in this class of occupancy is the tendency to overlook combustibles. Personnel occupying an office, as well as even some fire inspectors, may ignore the fuel potential in the paper accumulations in office buildings.

The office occupancy also has a higher degree of life hazard than is usually recognized. Although the density of population is not as great as in a public assembly

FIGURE 5.2 ◆ Industrial, institutional, and office occupancies need to have employees become familiar with fire alarm stations. *(Mead Westvaco, Luke, Maryland)*

occupancy, in many large office buildings the density is fairly close to that found in a school or hotel.

Consideration must also be given to the inclusion of other types of occupancies in a building that is primarily an office building. A restaurant and its inherent fire problems, for example, may present a life and fire hazard in a structure that would otherwise be considerably safer.

RESIDENTIAL OCCUPANCIES

The residential occupancy is where the largest number of fire fatalities occur. Generally, municipal fire department inspectors are not responsible for legal inspection of one- and two-family private dwellings or within individual apartments, although many cities provide for such inspections under a voluntary program. The fire department inspector is responsible for the inspection of public areas of larger residential occupancies, such as apartment houses, hotels, motels, dormitories, and rooming houses. The inspector and the owner/operator are responsible for making sure that people occupying such structures are housed under safe conditions. Multifamily structures occupied as condominiums pose a greater enforcement problem.

Hazards normally found in publicly occupied facilities are usually common ones. Likewise, fire causes generally fall into the same classification. The structures, by nature of their operation, contain combustible materials in the form of room furnishings. Causes are present in the form of smoking material, heating appliances, lights, television sets, and other electrical and heating appliances. It is impractical to eliminate these hazards and causes, although certainly electrical devices can be held to a safe limit.

FIGURE 5.3 ◆ Outside fire escapes are not considered as suitable means of egress for modern buildings.

◆ LEGAL AND MORAL RESPONSIBILITIES OF THE INSPECTOR

The inspector has a legal responsibility to check for all provisions contained in the code the agency is enforcing. Beyond this, however, is a moral responsibility to point out additional hazards or problems the inspector may note that are not covered under specific points of the applicable code. An example would be an item not specifically required for compliance because of a grandfather clause in the municipal charter or fire prevention code ordinance. The inspector may personally believe that conditions are dangerous, but because of the aforesaid provisions no retroactive application can be required. The inspector has at least a moral obligation to advise the owner or occupant of such unsafe conditions.

Subsequent to a major U.S. fire in which almost 100 people were killed, an examination of the most recent inspection report of the local fire department indicated that there were no violations in the facility. The inspection had been made three months prior to this tragic fire. At the time of that inspection, the inspector had observed conditions that eventually contributed to a major loss of life but failed to note them on the inspection report because of the existence of a grandfather clause. The inspector's attitude was that the problems were exempted by the grandfather clause, so it was not necessary to report them.

Grand jury investigations of this fire made public the fact that the building had received a clean fire safety report only a few months before the fire, yet the conditions that contributed to loss of life existed at the time of the inspection.

An interested citizen wrote a letter to the editor of the newspaper in the city after the grand jury findings had become public. The writer stated that she thought all municipal funds appropriated for fire prevention inspection purposes were "down the drain" if the inspection of this building in which the tragic fire occurred was an example of inspection procedures in the community. She made the point that if the inspections were so weak that no problems were reported in this building, the public could have little confidence in a fire prevention inspection conducted by the department. Although the inspector in question fully complied with all requirements imposed by code inspection procedures, mention of hazards not covered by the code would have been helpful.

The occupant of the facility that has been inspected may feel that the building is completely safe from a fire protection standpoint when the inspector leaves the premises or when an inspection report states that all deficiencies have been corrected. Generally, the inspector is looked on as an individual capable of providing complete and accurate information on fire spread possibilities, possible points of origin of a fire, and other related information.

Although many larger industries and institutions are inspected by representatives of fire insurance carriers, the fire service should not rely on these agencies to shoulder the responsibility of fire prevention. Inspections should be conducted and reported as if there were no other individuals having any responsibilities for making inspections of the place. The fire department inspector is probably the only one entering the premises who has a legal responsibility for conducting an inspection and taking steps to ensure compliance with code requirements.

All other fire safety inspections are advisory in nature; an insurance representative can speak only for the insurance company in determining the force that can be used in bringing about compliance. It is possible for the insured to seek insurance coverage from some other carrier if the inspection requirements are deemed too severe. As another alternative, the insured may elect to pay higher premiums to keep coverage in effect without making changes recommended by the insurance inspector or to merely accept the risk by canceling insurance coverage.

◆ CONFLICTS WITH OTHER AGENCY INSPECTIONS

Both in the exit interview and in the preparation of a report, it must be remembered that the property owner or operator is probably receiving fire-related recommendations from sources other than the fire prevention bureau. It is thus possible for conflicting recommendations or requirements to arise for the same deficiencies. The fire insurance representative might recommend one solution to a problem, while the fire department might recommend another. It is possible that the solution offered by the insurance inspector is as good as or even better than that developed by the fire department inspector. The latter individual should be willing to listen to other solutions to problems as developed by inspectors from insurance companies or by management itself. The inspector should not have a closed mind to solutions offered by others.

The fire prevention bureau inspector certainly has the final authority, although some other individuals inspecting facilities also may have legal authority in the fire prevention field. When management receives two conflicting legally constituted fire safety inspection reports, a difficult decision must be made. In such a situation, a conference of all interested parties is a method for resolving such a conflict. Another potential source of irritation to property owners and managers comes as a result of vague, nonspecific requirements being imposed by the inspector (i.e., fire extinguisher height "so as to be safely removed" as opposed to a specific mounting height).

Sometimes an inspector is discouraged at not receiving a warm welcome on arrival to make an inspection or from a lack of cooperation during the exit interview. This may not necessarily be attributed to a resentment of fire prevention inspectors. The cool reception may be blamed on the multitude of inspections to which the average industrial plant, mercantile establishment, or institution is subjected. In addition to fire safety inspections from public agencies, fire department prefire planning surveys, and sanitation inspections, the average mercantile establishment is also subjected to inspections for occupational safety, fire and casualty insurance, as well as other types of inspections. As many as 15 or 20 different inspection agencies, public and private, may call on a large industrial plant during the course of a year.

OCCUPANCY SELF-INSPECTION

In some occupancies, inspections may be carried out by inspectors employed on a full-time basis in the plant, institution, or mercantile establishment or by individuals stationed at the corporation's headquarters office. In the latter case, the inspector from the management level travels from plant to plant to conduct fire safety inspections. These inspections are carried out in much the same manner as those of the public fire service inspector, as the employee has the responsibility of assuring complete compliance with company fire safety regulations. They usually have a thorough knowledge of the business or industry involved, and, in fact, their specific knowledge of company processes often exceeds that of the public fire service inspector, who must try to learn about all types of occupancies.

Some fire departments use self-inspections for a selected, low-hazard group of occupancies. Arvada, Colorado, a district of 150,000 population, designated certain occupancies for this concept. By the use of computers, the department eliminated from the program those occupancies with high risks, hazardous materials, high exposures, fixed fire protection systems, and places of assembly. It then evaluated the remaining occupancies considering square footage and past fire safety code compliance problems. These reviews identified small businesses and light hazard warehouses as being the principal occupancies suitable for the self-inspection system.

A cover letter and legally correct self-inspection form was sent to the selected occupancies. Spot checks were made to ensure compliance. Some occupancies were scheduled for fire department inspection every other year, with a self-inspection the other year. A recent enhancement in career staffing resulted in Arvada returning to traditional inspection procedures.[3]

Potential legal ramifications must be considered before implementing a self-inspection program. The issue of liability for compliance failures due, for example, to misunderstanding the form is an important factor, as is the issue of the jurisdiction providing fire safety inspection services to some businesses and not to others.

CONSULTANT FIRE SAFETY INSPECTION

Some mercantile chains and larger industries engage the services of an outside consultant firm to carry out fire prevention duties. These firms assign personnel who are familiar with the given type of industry and have a considerable degree of familiarity with the processes or procedures of the organizations being inspected. If properly supported by the employing company, this type of program can be most effective.

The consultant concept is employed by a number of major concerns that are self-insurers. To be successful, a self-insurance program must include a provision for stern enforcement of fire protection requirements, either by company employees or through the use of an outside engineering firm.

INSURANCE CARRIER INSPECTION

The average mercantile establishment, institution, or industry is also faced with the possibility of an inspection by a representative of the insurance carrier. This inspection may be for the purpose of determining the insurance rates to apply to the property. Inspections may be conducted to determine possible limits of policy coverage and to tailor the policy to the specific situation encountered.

An insurance inspection may also be carried out for the purpose of rendering engineering services. This inspection may be by a representative of an individual insurance company or by a representative of an insurance group, such as the HSB Professional Loss Control or FM Global Group. The primary purposes of such an engineering inspection are to ensure that the fire protection equipment in the plant is in proper working condition and to identify any major hazards. Changes in procedure, layouts, construction, and protection are also noted.

The property owner will find that the inspection for engineering purposes includes testing fire protection equipment and available water supplies. This could include the tripping of dry pipe valves and flow tests on hydrants, tests of stationary fire pumps, and tests to ensure proper operation of all fire protection equipment in the structure. Although these inspections are most helpful, they are an interruption to the routine of the plant. Management must provide personnel to carry out the duties of opening valves, resetting dry valves, etc., as a part of the tests.

OTHER GOVERNMENT AGENCY INSPECTIONS

In addition to the previously mentioned private inspections, the industrial plant or mercantile establishment will from time to time be inspected by other governmental agencies. The health department, for example, will probably inspect the facility for overall sanitation and for food handling procedures. Institutions will likewise be inspected by the health department to determine compliance with standards relating to nursing care, patient health, administering drugs, etc.

The labor department or department of labor and industry in practically every state has a responsibility for employee safety. These inspections may be conducted by the U.S. Department of Labor inspector. As mentioned in Chapter 10, this responsibility may cover the provisions of the Occupational Safety and Health Act of 1970. The safety inspector will usually take immediate action on conditions encountered that affect the safety and well-being of the employees, including basic fire prevention items.

The local industrial plant may have prefire planning inspections by personnel from fire fighting companies separate from fire prevention inspection activities. Regardless of coordination and timing, personnel from the plant should take the time to accompany the inspectors in carrying out these duties.

The federal government may send inspectors into certain plants through the provisions of the Port Security and Waterways Act of 1972. This act provides for Coast Guard inspection, regulation, and enforcement in facilities adjacent to navigable waterways. The Coast Guard is empowered to bring about compliance, which includes fire safety, in such facilities.

A federal inspector may likewise inspect a building for fire protection purposes if the plant has contracts for providing services or materials to federal agencies or has processes subject to federal inspection, such as food preparation. These inspections are designed to ensure the continued or safe production of products under review and may include coverage of fire safety items. In some plants, an inspector may be kept on the premises at all times.

Although all of the preceding inspections are necessary, they consume a considerable amount of management's time as well as that of maintenance personnel. Plant management may not welcome the fire inspector because of the multitude of inspections. Time taken by plant personnel in accompanying the inspector and by management in discussing requirements is time away from production.

◆ **SUMMARY**

A key function in the enforcement of fire laws and regulations under the jurisdiction of the fire department is the fire safety inspection. Besides having a thorough knowledge of code requirements and competency in inspection procedures, the fire safety inspector should have good judgment, keen observation, and skill in dealing with people.

Before going out on an inspection, the inspector should be prepared with the proper tools: a flashlight, camera, notebook, data logger, recorder, or computer (personal data assistant) for taking notes on the inspection tour; code books and handbooks should be available in the car for reference. The inspector should also look over any reports of previous inspections of the facility for background information to enable proper follow-up on earlier recommendations. Under no conditions should an inspection be undertaken when the inspector is pressed for time and unable to make a thorough inspection.

On arrival at the site, identification must be presented to a person in authority, and permission must be secured to make the inspection. This is a legal obligation as well as a courtesy. Court cases have challenged the legality of inspections made without proper identification and permission. It is also customary for the inspector to be accompanied by a representative of management.

The tour usually starts at the top of the structure with an inspection of the roof, then proceeds to the top floor and on down through the entire structure, including the basement levels. A systematic inspection should be made of all spaces on every floor. If access to certain spaces is not immediately possible, the inspector must wait until permission for entry to these can be arranged.

The objective of a fire prevention inspection is to discover hazards and make recommendations or impose requirements to abate or eliminate them to the greatest extent possible. Conditions to look for are exposure hazards, potential fire hazards, exit deficiencies, or any unusual conditions. Besides noting the obvious potential fire causes or storage of flammable liquids or gases, the competent inspector should look for latent hazards, such as ignition sources from operating equipment, and should be able to recognize fire spread possibilities inherent in the design and construction of the building. Under no conditions should the fire inspector leave the premises without being sure that all fixed fire protection equipment is in proper functioning order as required by applicable fire prevention codes. And aside from the mandatory responsibility to ensure compliance with legal requirements of the code, there is a moral obligation to point out to property owners any potential hazards not specifically named in the code.

As many easily correctable violations as possible should be taken care of during the inspector's visit. Changes necessary for compliance with the fire code should be discussed with management at the end of the inspection tour. Recommendations should be clear and concise and should be carefully weighed to make sure they will correct the situation and not put the owner to needless expense.

A report should be made of all fire prevention inspections to give the owner notice of violations and to have a record on file for follow-up of recommended corrections. This report would also be used for reference in possible legal action to enforce compliance. An owner is usually given a specific time in which to complete the requirements. If reinspection finds the work not completed, a legal order is issued to comply or face a possible fine or jail sentence.

Among general considerations, the fire safety inspector should be aware of hazards peculiar to various types of occupancies, such as places of public assembly, institutions, mercantile establishments, warehouses, office buildings, hotels, and residential occupancies. For example, such hazards include blocked exits or passageways in public assembly places; grease in restaurants; and the hazards of fire in institutions with bed patients or older adult residents. The inspector should also be aware of possible conflicts with inspections carried out by insurance carriers and other local, state, and federal agencies and with industry self-inspections and municipal grading schedule surveys. Because the various inspections a property is subject to may put an owner in a quandary, the fire inspector should be prepared to cooperate in working toward an accommodation of all interests without mitigating fire safety.

Review Questions

1. Detailed information regarding inspections can be found from the NFPA or
   a. IFSTA
   b. IAFC
   c. IAFF
   d. ISFSI
   e. IAAI
2. Inspection frequency is based upon all of the following except
   a. availability of personnel
   b. degree of hazard in the occupancy
   c. potential for finding deficiencies
   d. fire suppression forces
3. Before conducting an inspection, the following should be done.
   a. Review past reports
   b. Confer with previous inspector
   c. Review code book based upon occupancy
   d. All of the above
   e. None of the above
4. Inspectors should proceed with a routine inspection only if they
   a. have permission from the proper authority
   b. are certified
   c. are ordered by a superior
   d. feel like it
5. Combustibles, if ignited, that would cause the spread of fire that should be checked by inspectors include all but
   a. flammable liquids
   b. oxidizing agents
   c. explosives
   d. acids
   e. metal
6. Sources of ignitions to be considered during inspections include all but the following.
   a. fuel leaks
   b. electrical arcs
   c. open flames
   d. smoking
   e. static electricity
7. Absence of fire division walls and improper fire doors contribute to
   a. vertical fire spread
   b. horizontal fire spread
   c. lateral fire spread
   d. explosions
   e. flashover
8. All exit interviews must be followed by
   a. a phone call
   b. review by superior
   c. a written report
   d. court summons
9. Time allowed for the anticipated completion of fire code violations is
   a. always short
   b. dependent upon the complexity of the violations
   c. always the same
   d. none of the above
10. Fire prevention inspectors' results may conflict with
    a. fire consultants' inspections
    b. insurance carriers' inspections
    c. other government agency inspections
    d. all of the above
    e. none of the above

Answers

1. a
2. d
3. d
4. a
5. e
6. a
7. b
8. c
9. b
10. d

■■

Notes

1. Amendment 107, *Volume V Fire Prevention Code* (Raleigh, N.C.: North Carolina Building Code Council, 1991).
2. *City of Toledo v. Seto, Inc.* (1996, Ohio)
3. Personal communication to the author from Kent Shriver, Arvada Fire Marshal, 2004.

Preparing Fire Service Personnel for Fire Prevention Duties

6 CHAPTER

The initial fire protection emphasis by many Colonial era communities was in fire prevention through the appointment of a fire warden. However, the primary motivation soon became that of suppression with the majority of post–Colonial era communities starting fire services as the result of a major fire.

Many people believe that the fire department's obligations have been met if the department responds to and brings under control all fires to which it is called. Fire prevention in the past has not been generally thought of as being a part of the basic responsibility of many fire departments. Fortunately, the importance and effectiveness of fire prevention has been acknowledged by the fire service and the public, and prevention is once again becoming a major fire service function.

◆ LACK OF EMPHASIS ON FIRE PREVENTION

Training programs in the fire service have likewise had limited emphasis of fire prevention. They have for the most part concentrated on fire suppression subjects. The average individual joining either a career, volunteer, or combination-type fire department has received little information relating to fire prevention in the basic training program. Members of many explorer and other Scout groups receive more training in fire prevention than does the person joining the average fire department. Fortunately, the trend in the fire service has been to include fire prevention training as part of the basic training program. Several programs directed at the volunteer fire service also include fire prevention training.

Major fire conferences designed primarily for fire service audiences have not been noted for their strong emphasis on fire prevention. Talks on fire prevention have not been found on fire service conference programs to any great extent.

National periodicals in the fire service field have likewise given a lesser amount of coverage to fire prevention than to other phases of fire service activities. A change in emphasis is beginning to appear, and periodicals are paying more attention to fire prevention.

There are probably other reasons than those already mentioned why fire prevention has not been well recognized in the overall fire service picture. Motivation is certainly more limited because of the less glamorous nature of the tasks performed. The excitement and activity of the fire scene are far more challenging and rewarding to many people than is performing a building fire prevention inspection or staging a demonstration before a group of 6-year-olds.

The fruits of labor are also much more readily recognizable in fire suppression efforts than in fire prevention work. The individual engaged in fire suppression activities is rewarded with direct evidence of something accomplished. The control and extinguishment of a major fire are highly visible results, whereas the rewards of fire prevention work are usually intangible and not so easily measured.

An individual assigned to fire suppression and emergency medical duties seldom comes in contact with the public in adverse situations. The fire preventionist is often in a confrontational relationship with others, especially in code enforcement. Citizens do not usually care for advice to extinguish cigarettes, provide automatic sprinkler systems, remove obstructions from exits, and complete other fire safety requirements. They are, in fact, often alienated by such efforts. This is probably another reason for the reluctance by some fire service personnel to accept fire prevention as a major fire department activity.

Some fire service personnel believe that the insurance industry is the major benefactor from any efforts they might expend in the fire prevention field. Such a thought is probably based on the fallacious assumption that insurance companies directly profit from a reduction in fire losses. In fact, insurance rates are adjusted according to actual losses.

◆ FIRE PREVENTION TRAINING FOR ALL FIRE SERVICE PERSONNEL

In preparing fire service personnel for fire prevention duties, it is important to consider the needs of the individual firefighter. We should consider the motivation factors.

One important incentive is that fire safety work promotes the professionalization of the fire service because of the higher level of knowledge and skills required to do the job. There is no question that certain skills required for fire prevention assignments are more advanced than those required for basic fire suppression duties. Technical training and proficiency are needed to understand the hazards of industrial processes and the uses of dangerous chemicals and to evaluate construction methods. Fire prevention training opens up new fields of specialization for fire service personnel and widens their range of activities in the service.

The firefighter also gains knowledge that is helpful in fire suppression duties. In conducting fire inspections and prefire planning surveys, the firefighter has an opportunity to learn about conditions in the buildings in the community before a fire occurs. This information can be of great help when entering a building during a fire emergency.

A vigorous fire safety program that places fire service personnel in frequent contact with the people of the community can do much to enhance the image of the fire department. It may help to erase some preconceived notions on the part of the

Figure 6.1 ◆ Rapt attention of young children to the firefighter's message is not unusual. *(Grand Lake, Minnesota, Fire Department)*

public of the work of the fire service. Too often, the picture of the firefighter in the public mind is that of an individual leaning back in a chair in front of the fire station, "sitting and waiting for a fire." If the public can be brought around to thinking of the firefighter as an active public service employee who is on the street preventing fires when not engaged in emergency medical or fire fighting activities, certainly, the overall good is served.

Public contacts by fire service personnel in fire prevention duties, especially if they are favorable in nature, have other benefits, as citizens become more aware of the hazards of fire and gain confidence in the fire service, they may be more inclined to call the fire department without delay in the event of an incipient fire, no matter how small, instead of trying to cope with it themselves and risking serious injury or greater property loss. Many communities have found that after a stepped-up fire safety program was put into operation, the number of fire calls was actually reduced, presumably because fire prevention advice was being followed. There is less likelihood that this would have a major effect on the loss per fire, but any reduction in the number of fires means an overall gain to the community.

Because of the community involvement aspects of fire prevention work and the opportunities they provide to build good public relations, public respect for the fire service will increase. This can have a favorable effect on the ability of the fire department to obtain funding for needed apparatus, personnel, and new stations. The citizen who is favorably impressed with the fire service is ready to support efforts to increase the department's personnel, improve facilities, and otherwise upgrade the services rendered. Provision of emergency medical service also has this effect.

◆ MASTER PLANS FOR FIRE PROTECTION

In order to successfully prepare fire service personnel for fire prevention duties, it is well to first consider the entire fire protection picture in the community. Individuals undertaking this work will have a much better understanding of the importance of their roles if they are aware of the concepts embodied in a fire protection master plan.

America Burning, the report of the National Commission on Fire Prevention and Control, contained a recommendation stating that "The Commission recommends that every local fire jurisdiction prepare a master plan designed to meet the community's present and future needs in fire protection, to serve as a basis for program budgeting, and to identify and implement the optimum cost-benefit solutions in fire protection."[1] Legislation enacted pursuant to this report also strongly encouraged the promotion of the planning process and defined a fire defense master plan as "—one which will result in the planning and implementation in the area involved of a general program of action for fire prevention and control. Such a master plan is reasonably expected to include: (1) a survey of the resources and personnel of existing fire services and an analysis of the effectiveness of the fire and building codes in such area; (2) an analysis of short- and long-term fire prevention and control needs in such area; (3) a plan to meet the fire prevention and control needs in such area; and (4) an estimate of cost and realistic plans for financing the implementation of the plan and operation on a continuing basis and a summary of problems that are anticipated in implementing such plan."[2]

This concept envisions that the entire fire problem in a community be given comprehensive review and that a plan be prepared and implemented to address the needs of the jurisdiction. Fire prevention, including fire and building code enforcement, zoning provisions, fire safety education, installation of fixed fire protection in structures, and a myriad of other fire protection–related areas would be considered in the plan. The development of the plan, known as a master plan, must involve community leaders, not just fire service personnel, in order to be successful. Costs are a key element.

A master plan results in an organized means of defining and implementing a level of fire protection in a community that equates with the desires of the citizens, coupled with legally mandated requirements. A comprehensive database including demographics, property evaluations, fire experience, and several other elements is needed to develop and implement a plan. Acceptable life and property risk levels should be established in the plan.

Direct and indirect costs of fire protection need to be considered. For example, many businesses damaged or destroyed by fire do not rebuild in the same community, thus creating a loss of tax base. The cost of treatment of burn victims is another example.

Although the cooperation and assistance of a wide variety of agencies and organizations is necessary in order to prepare and execute a master plan, the fire department is always the primary instigator and the agency that "rides herd" to be sure the job is completed.

Those fire service personnel engaged primarily in fire prevention work seldom are in the limelight for their efforts. An appreciation of the community's master plan will enhance their personal pride in the major role they play in the safety of their fellow citizens.

Howard D. Tipton, then administrator of the National Fire Prevention and Control Administration and the recognized instigator of master planning for the fire service, provided this summary of the concept:

The plan provides an organized means to document the current fire protection elements and services provided to the community and the future service required by the community in order to achieve the goals and objectives. Definition of the basic elements of fire protection (e.g., water supply, zoning, suppression, prevention, life safety, etc.) and establishment of desired types and levels of services provide a basis for community awareness of the extent of the services offered. Further, the community is made aware of the risks involved and the degree of private sector participation required to complement the public sector services for meeting the goals and objectives.[3]

◆ RECRUIT TRAINING PROGRAMS

An example of an excellent training program in the field of fire prevention is that of the Dallas, Texas, Fire Department. As part of the recruit training school, each new member is required to undergo an additional 56 hours of training in fire prevention and related subjects.

The fire prevention subjects are designed to give the recruit an appreciation of the worthwhile purposes of fire prevention duties. Subjects covered during the fire prevention phase of the training program include objectives of fire department inspections, fire hazards and causes, inspection techniques and procedures, prefire planning and inspection, fire cause determination, and fire protection systems.

The New York City Fire Department likewise has a program for indoctrinating new personnel in fire prevention methods. It includes fire prevention training at the fire academy and in the station as part of its overall program. The recruit is teamed with a company officer during apparatus field inspections and is given the opportunity to apply fire prevention principles he or she has learned. This hands-on training with the officer allows any weaknesses to be revealed and later strengthened through additional fire prevention drills and discussion.[4]

Louisville, Kentucky, includes 80 hours of class on fire prevention and public education in its 720-hour recruit training program. Subjects include methods of conducting home and nonresidential building inspections, technical fire codes, home fire safety lessons, and public fire education. As a final examination, recruits are required to conduct fire safety classes for high-risk groups in the city.[5]

Many cities assign fire suppression companies to conduct code enforcement inspections. Training must be provided to participating fire company personnel. Some states require instructors as well as inspectors to be certified. In Salina, Kansas, for example, fire prevention officers are authorized by the state to serve as instructors. They, in turn, instruct all company personnel in basic fire prevention inspection and cause and origin determination. All firefighters and company officers serve as inspectors.

◆ PREFIRE PLANNING

There is a very close relationship between prefire planning and structural control inspections. Some fire departments have attempted to combine these visits; however, this is not advisable because of the differing purposes of the activities.

Prefire planning surveys are primarily carried out for the purpose of familiarizing fire department personnel with building locations and floor plans as well as sources of potential danger or other unusual features. The surveys also provide fire suppression personnel with knowledge of the locations of sprinkler control valves, standpipe connections, and other information relating to the fixed fire protection equipment in the building. The location of hazardous commodities that may endanger fire fighting personnel is also noted. Information obtained in a prefire planning survey must be available for personnel responding to emergencies. No doubt many firefighters have avoided death and injury because of prefire planning programs.

Prefire planning surveys and fire prevention inspections are devised for two different purposes, and combining the two into one inspection may place the property owner in a rather difficult position. The owner is interested in providing fire suppression personnel with information helpful in fighting a fire but is probably not as interested in having fire personnel enter for the purpose of bringing about corrective action. If the same firefighters are assigned to carry out both responsibilities at the same time, there is a possibility that the two undertakings will conflict with each other.

In some jurisdictions the word *planning* has been deleted from the title of this program because of potential legal liability in the event that fire fighting efforts are unsuccessful despite the presence of a "plan." In other locations the term *prefire survey* is used; however, this may cause problems because it presumes that there will be a fire. Regardless of the title used, this training of personnel is imperative.

◆ FIRE PREVENTION TRAINING ASSIGNMENTS

Another approach is to assign personnel to fire prevention functions at some time during their career in the fire department. In several departments, for example, personnel promoted to the rank of lieutenant must, during their tenure in this rank, serve in the fire prevention bureau for a specified period of time. This procedure has been found to be quite helpful. An individual moving up through the rank structure gains an appreciation of fire prevention. These practices do a great deal to preclude the possibility of an individual reaching the position of chief officer without having served in fire prevention.

Greater emphasis is being given to the use of all active fire service personnel for fire prevention duties. A time may come when fire prevention will be a recognized duty assignment for anyone entering the fire service, as is the case with forest fire fighters. Such an approach will go a long way toward reducing fire losses.

Georgia's fire commissioner appointed a 60-person commission to study ways to reduce injuries, deaths, and property damage caused by fire. Members representing both public and private sectors held hearings in 12 Georgia cities to obtain input. The Commission was divided into four committees: Training and Operations, Codes, Investigations, and Public Education. Their final report was issued late in 1993.

A major recommendation of the Codes Committee was that provisions be made to permit building officials to be trained at Georgia's Public Safety Training Center, which is the state's fire training center. The report points out the advantage of having joint training of building and fire officials.

The Public Education Committee recommended that public fire safety education be made a part of basic minimum standards training for both career and volunteer firefighters. They also recommended that newly elected local public officials, as well as newly appointed fire chiefs, receive indoctrination in public fire safety education.[6] New firefighters now receive this training.

Many states have training programs to prepare persons to conduct fire safety education programs. Several states now have certification provisions for these individuals.

In addition to the indoctrination and training of general fire service personnel in fire safety duties, there is a need for training in inspection procedures for full-time personnel in the fire prevention bureau.

As an example of early training in fire prevention duties, the following instructions were given to fire inspectors in the Philadelphia Fire Department in 1913:

> Major fundamental principles in the training of fire inspectors:
>
> 1. The work of fire prevention may impress you as a fad or a joke at the outstart, but your matured view will be that it is both interesting and serious and in comparison with fire extinguishing it stands as a Bureau of Police to the Detective Bureau.
> 2. Your sound judgement of corrections to remove fire dangers still has the backing of your superior officers against all protests, and to secure the desired result, you are to permit no hindrance in your work through fear or favor, friends or enemies.
> 3. Be courteous, explain in detail your mission and do not quarrel with the occupants of any building. Remember there is a distinct advantage over the man who has lost his temper to the man in an argument with self-control. You are representing the dignity of a department of your city and must in no case initiate a quarrel; if you cannot, report the facts to the officers in charge of the fire prevention work and have them open this avenue for you.
> 4. Make friends with the engineer, janitor, superintendent or any other person in charge of the building. These men have considerable information of the property condition and influence in having corrected promptly conditions which you desire improved.
> 5. This new responsibility calls for intelligence, sound judgement, broad views and being amenable to reason. Narrow-mindedness and bombast have absolutely no place in this work and every new and practical idea bearing on the end ultimately to be achieved should be promptly grasped and utilized.
> 6. Gently and firmly decline the friendly cigar or other small article usually offered in good faith, but occasionally used by occupant or owner of property to put the inspector in a bad light. In particular, guard against the least suspicion or suggestion of graft in this form and by adhering to strictly business dealings you are in a stronger position should your work be questioned or reviewed.[7]

These principles are as applicable today as they were in 1913. They should be considered and adhered to by everyone assigned to fire inspection duties.

The more an inspector knows, the more valuable are the inspections. A major problem in effective fire prevention inspection is that of failing to see the significance of what is being inspected. This failure may be the result of lack of training in fire safety inspection, or it may simply be the individual's inability to register problems seen during the inspection. An inspector may report that a complete inspection has been made by going around and checking on fire extinguishers and on enforcement of smoking regulations, when in reality an effective inspection has not been made.

Educational needs include training in inspection procedures, familiarity with fire and life safety codes, training in plan reviews, an appreciation of building code fire protection provisions, an understanding of health codes as related to fire safety, and an understanding of court decisions affecting the field. The fire inspector should also have a general knowledge of procedures to determine the causes of fires.

In smaller fire departments, where it may not be practical to establish formal programs, training procedures often consist entirely of on-the-job training. With this method it is essential that the trainee be assigned to work with an inspector who is fully qualified to perform inspection duties. Such training might take place in another city with a larger bureau, or at a state fire academy. The National Fire Academy also offers extensive training in this field.

Many larger departments have formal training programs for fire safety inspectors. The case study method is often employed in these programs. Case histories of actual field problems in fire prevention and fire code enforcement are studied and analyzed as part of the training program.

The fire safety inspector should have available reading material pertaining to changes in the field. All publications, including books and magazines relating to new processes, new fire protection equipment, etc., should be routed through fire prevention inspectors. Weekly conferences and training sessions are also helpful.

Although technical fire inspection duties require more extensive training, a great deal of effort is not always necessary to prepare fire service personnel for basic fire prevention duties. Some programs, including home inspection and basic fire prevention education, can be made effective with relatively simple training methods.

FIGURE 6.2 ◆ The widespread inclusion of EMS as a fire service function has expanded the duties of fire department public educators. *(Photo of Camden, Maine, Fire Station by Don Campbell)*

◆ NATIONAL PROFESSIONAL QUALIFICATIONS SYSTEM

The Joint Council of National Fire Service Organizations, a group composed of the major national fire service organizations, agreed in 1972 to establish standards for professional competency in the fire service. The Joint Council no longer exists. Committees were formed to develop and prepare recommended minimum standards of professional competence for four groups: firefighters, fire inspectors and investigators, fire service instructors, and fire service officers. The composition of these committees was based on Joint Council membership.

The National Professional Qualification Board was established by the Joint Council of National Fire Service Organizations to oversee and validate the standards developed by the four committees. The board was also responsible for establishing and maintaining a nationally coordinated professional development program for the fire service. In 1990 responsibility for oversight of the Professional Qualifications project was assumed by the National Fire Protection Association. The committees are established and operated under the National Fire Protection Association's standards-making procedures, and proposed standards and revisions are submitted to the Association for final adoption.

The standards may be used by states and communities as part of their requirements. The system is designed to encompass all duties of the uniformed fire service member and certain "civilian" positions.

It was proposed that each state have a certification process. The National Board on Fire Service Professional Qualifications accredits state certification boards and other agencies that certify persons to NFPA standards. The International Fire Service Accreditation Congress also accredits state programs. Certification at each level, starting with Fire Fighter I, would result from the individuals successfully completing certain performance objectives. The individual's progress is based on the ability to master requirements and demonstrate proficiency to an examiner. Of course, promotions would be limited by the job vacancies in the fire department.

The professional qualifications system does not attempt to assign ranks to levels of qualifications. It envisions that local jurisdictions would make these determinations. A wide range of interests are represented by the National Fire Protection Association committees, which develop the standards.

The term *professional* in this context connotes individual abilities rather than employment status. Volunteer and combination career and volunteer departments can participate in the system; in fact, many are already utilizing the standards.

The system envisions that training programs will prepare fire service personnel to acquire skills and knowledge necessary to achieve the performance objectives of the standards. It also envisions a controlled independent testing procedure for candidates.

The standard, *Professional Qualifications for Fire Inspector,* National Fire Protection Association (NFPA) No. 1031, was designed to ensure the competency of personnel assigned to these duties.[8] The standard is not designed to be used for personnel other than those in the fire service. It does include "direct entry" by individuals who are not firefighters, however.

The updated procedure as approved in 1987 divides the original standard into three standards, NFPA 1031 (*Professional Qualifications for Fire Inspector*); NFPA 1033 (*Professional Qualifications for Fire Investigator*); and NFPA 1035 (*Professional Qualifications for Public Fire and Life Safety Educator*). This arrangement provides qualification requirements for individuals entering directly into the three specialist areas. The growing tendency to employ individuals specifically for assignments in fire inspection, investigation, and public fire safety education brought about this change.

Full use of the fire inspector, fire investigator, and public fire educator job descriptions is found only in a larger fire department. Hundreds of smaller fire departments assign only one person to fire prevention, investigation, and education duties. In these communities, the one individual would need to meet the requirements for all of the duties imposed.

Two of the three duty assignments may be combined in some fire departments; that is, the same person may carry out inspection and investigation duties while another person performs education duties. In a number of communities, inspection and education duties are combined with a separate person or persons performing investigation duties. In either situation the individuals so assigned would need to master the requirements for their duties. The standards do not envision that a community will necessarily establish separate positions for each job classification or level of progression. Some communities or states may desire to impose higher standards than those included in the NFPA standards.

Candidates for progression under the standards are required to demonstrate an ability to express themselves clearly orally and in writing. They are also required to demonstrate their ability to interact with the public under scenarios of code enforcement, fire investigation, or fire safety education duties with tact and discretion. Candidates are also required to demonstrate knowledge of personal safety practices and procedures.

People aspiring to be fire inspectors at level I must demonstrate knowledge and abilities in many specific areas, including flammable and combustible liquids, compressed and liquefied gases, explosives, heating and cooking equipment, principles of electricity, safety to life, code enforcement procedures, fire cause determination, and miscellaneous provisions. The fire inspector II is required to master these subjects in greater complexity and to be able to develop plans relating to fire safety. As previously mentioned, the fire inspector III must also master administration and management functions.

In states where fire investigators are considered to be officers of the law, they are required to meet applicable minimum standards set for law enforcement officers. They must also demonstrate knowledge and abilities in areas including legal, evidence collection, photography, fire scene examination, records, reports and documents, and courtroom procedures.

The public fire and life safety educator must demonstrate knowledge and abilities in general areas and in public information and public education fields. The individual must have teaching abilities in order to fulfill these requirements successfully. He or she must also understand plant and institutional emergency organization and have associated instructional abilities. The public fire educator III also must master administration and management functions, including an ability to develop examinations and evaluate personnel skills.

As a specific example, the fire inspector I must be able to compute the allowable occupancy load of a single-use building or portion thereof, given a detailed description of the building or portion of the building, so that the calculated allowable occupant load is established in accordance with applicable codes and standards. The public fire and life safety educator III must create a program budget given organizational goals, community needs, and budget guidelines, so that overall program needs are met within budget guidelines.

There are many reasons for the adoption of professional qualification requirements for fire safety personnel. The system helps develop qualified personnel and goes a long way toward assuring the city administration and the public that fire prevention, investigation, and education duties are being carried out by qualified individuals.

Several state court decisions provide an impelling reason for ensuring that personnel are fully qualified. The necessity for courtroom qualification of fire investigators as expert witnesses makes the system almost imperative for these people.

A growing number of states are requiring individuals who perform fire safety inspections to be certified under a state program. Some states also certify fire investigators and public fire and life safety educators. In these states the NFPA standards are generally the basis for the certification requirement.

Many jurisdictions use fire suppression personnel for fire inspection duties. In states with certification requirements, this means that all firefighters must be certified as fire inspectors. Firefighters are carrying out the majority of fire inspections, and fire prevention officers serve as resource specialists and handle complex fire safety assignments.

Many other fields of endeavor have had professional qualification systems in effect for years. These include public accountants, insurance underwriters, electricians, barbers, beauticians, and a variety of other employment categories. The public can have reasonable assurance that a barber is adequately trained but cannot always have that assurance with a fire inspector. Under the professional qualifications system, a fire inspector I in New Hampshire has basically the same abilities as a fire inspector I in Arizona or New Jersey.

Over 220 two-year fire science programs may be found in community colleges in the United States. Many of these programs offer courses in fire prevention and fire investigation. Attendance is primarily from public fire departments; however, building inspectors, architects, and engineers may be found in these classes as well. Most fire science programs offer courses in building and fire prevention codes and standards.

Many fire departments are encouraging fire prevention bureau personnel to enroll in fire science programs. Time off to attend classes or special financial stipends may be offered as inducements for enrollment.

The four-year fire protection and safety engineering technology program at Oklahoma State University, Stillwater, has been looked on as a leader in academic fire protection programs. Since the program's inception in 1937, hundreds of graduates have had a substantial impact on fire prevention and fire safety in many countries. Graduates obtain a bachelor of science degree in Fire Protection and Safety Technology.

There are 26 four-year fire science/fire protection–related program in the United States. Seven of these programs are affiliated with the *Degree at a Distance Program* sponsored by the U.S. Fire Administration/National Fire Academy. In addition, the University of Florida has an Internet-based program that also provides courses in EMS and Emergency Preparedness Management and Planning. Several offer advanced degrees. The University of Maryland, College Park, is viewed as a leader in offering fire protection as an engineering curriculum.[9]

◆ SUMMARY

The historic primary motivation behind the organization of most fire departments has been the suppression of fire in the interest of public safety. Little thought has been given to the allied function of fire prevention. Training programs for fire service personnel have concentrated for the most part on fire suppression activities, and there has been a conspicuous lack of emphasis on the prevention of fire, even in the programs of the major fire service conferences and in the national periodicals that serve the fire protection field.

A recent trend has been toward recognition of the rightful place of fire prevention as a twin function of fire protection. Major fire departments are beginning to include fire prevention training as part of their basic training programs. More attention is also being paid to the subject of fire prevention by writers and periodicals in the field, and many city charters specifically name fire prevention as a public responsibility.

Fire prevention training for all fire service personnel can have beneficial results in many ways. It promotes professionalization of the fire service because more knowledge is required than for basic fire suppression duties, and it opens up new fields of specialization for growth on the job for all fire service personnel. It also helps to generate respect for the goals of fire prevention.

A vigorous fire prevention program can also do much to enhance the image of the fire department in the community. If the public recognizes fire service personnel as active public servants who are on the street preventing fires when not engaged in fire fighting or emergency medical activities, certainly, the overall good is served. The citizen who is favorably impressed with the fire service is more inclined to support efforts to obtain funding for needed apparatus, equipment, new stations, and other fire department improvements.

Among fire departments that have introduced fire prevention subjects in their recruit training programs are the Dallas, Texas; Louisville, Kentucky; and New York City departments.

Specialized training needs for fire prevention inspectors include a thorough comprehension of the local fire code and procedures used to conduct an effective inspection; an understanding of other related codes, such as building codes and health codes; familiarity with fire investigation procedures; and recognition of the legal aspects of inspection responsibilities. Many of the larger departments have formal training programs for fire prevention inspectors; smaller departments may employ on-the-job or state training. One of the earliest training programs for fire inspectors on record is a 1913 program of the Philadelphia Fire Department, which outlined principles that are surprisingly applicable today.

Although the more technical fire inspection duties require more extensive training, basic fire safety education and home inspections can be made effective with relatively simple training methods.

An important development was the establishment of a National Professional Qualifications System for state certification of fire service personnel at several levels of professional competence in the classifications of firefighter, fire inspector, fire investigator, public fire and life safety educator, fire service instructor, and fire service officer. Standards were originally developed by committees appointed by the Joint Council of National Fire Service Organizations with staff support by the National Fire Protection Association. Adoption of these standards is a major step in professionalizing the fire service and improving the nation's fire protection capabilities.

Review Questions

1. Fire prevention has not always been well recognized in the overall fire service picture because
   a. fire suppression is more exciting
   b. fire prevention is rewarded with immediate evidence
   c. fire suppression is confrontational
   d. there are more national fire prevention conferences
2. Fire prevention and suppression
   a. are best separated
   b. are equally important
   c. are not related
   d. require different personnel

3. Master plans should include
   a. survey of resources and personnel
   b. short-/long-term prevention and control needs
   c. a plan to meet needs
   d. estimate of cost
   e. all of the above
4. The purpose of a prefire plan is to
   a. preplace apparatus
   b. establish command strategies
   c. familiarize personnel with building locations and floor plans
   d. all of the above
   e. none of the above
5. Fire inspectors can gain training
   a. at the National Fire Academy
   b. from state organizations

c. in larger departments
d. all of the above
e. none of the above
6. NFPA Standard _____ applies to fire inspectors.
   a. 1001
   b. 1021
   c. 1031
   d. 1041
   e. 1061
7. Additional NFPA standards for fire prevention include
   a. 1901 and 1902
   b. 1033 and 1035
   c. 1500 and 1561
   d. 13 and 231
   e. 10 and 20

Answers

1. a
2. b
3. e
4. d
5. d
6. c
7. b

Notes

1. U.S. National Commission on Fire Prevention and Control Report, *America Burning* (Washington, D.C.: U.S. Government Printing Office, 1980), p. 25.
2. Public Law 93-498 Federal Fire Prevention and Control Act.
3. Howard D. Tipton, *Master Planning for Community Fire Protection,* Annual Meeting, International City Management (Toronto, Canada, 1976), p. 8.
4. Personal communication to the author from Edmund P. Cunningham, Chief of Fire Prevention, New York Fire Department, 1998.
5. Russell E. Sanders, "Chiefs Briefing," *NFPA Journal,* National Fire Protection Association (November/December, 1993), p. 14.
6. Commission for a Fire Safe Georgia, *Goals for a Fire Safe Georgia* (Atlanta, Ga.: Office of Safety Fire Commissioner, 1993), pp. 13, 22.
7. Powell Evans, comp., *Official Record of the First American National Fire Prevention Convention* (Philadelphia: Merchant and Evans Co., 1914), p. 150.
8. *Professional Qualifications for Fire Inspector,* NFPA No. 1031.
9. Personal communication to the author from L. Charles Smeby Jr., Academic Instructor, Florida State Fire College and University of Florida.

CHAPTER 7

Organization and Administration of Municipal Fire Prevention Units

F ire prevention has been recognized as a function of municipal government since the earliest days of this country, as discussed in Chapter 1. In the Colonial period, the administrative functions of the town with respect to fire, police, sewage, water, health, and street lighting were often carried out through volunteer citizen service. Some of the early municipal charters required citizens to serve in certain positions of responsibility without pay, and people usually readily accepted such responsibilities. Voluntary fire protection and fire prevention service is provided in thousands of communities in the United States.

◆ THE PLACE OF FIRE PREVENTION IN MUNICIPAL GOVERNMENT

By tradition, fire prevention and fire protection in the United States have been primarily local government responsibilities. Fire and fire-related problems are more locally confined than are most other concerns of public services, such as crime, health, welfare, and employment. It is quite unusual, for example, for a structural fire to actually burn in more than one jurisdiction. Furthermore, fire safety code enforcement can usually be most effectively carried out at the local level.

The U.S. Fire Administration report *America at Risk* amplifies this concept:

> The primary responsibility for fire prevention and suppression and action with respect to other hazards dealt with by the fire services properly rests with the state and local governments. Nevertheless, a substantial role exists for the federal government in funding and technical support.[1]

In most communities a fire department is usually formed when something happens to demonstrate the need for fire protection for public safety and welfare. A community typically gets along without a fire department until it becomes obvious to the citizens that they can no longer rely on neighboring communities to provide fire protection or until a large fire involving loss of life or serious property damage brings urgent demands for community action to organize a fire department.

In most areas of the United States, the incorporated municipality is looked to as the most logical agency to develop and operate a fire department. In some areas, fire protection is not considered a municipal function, and service is provided by an incorporated fire department. This arrangement is often used in unincorporated communities.

In municipal government, elected officials are responsible for the overall administration of government. Administrative heads of departments, who are responsible for carrying out the day-to-day functions of the government, may be political appointees or may be appointed through a merit system. Lower ranks of municipal employees are usually selected through merit system or civil service procedures.

Recognition of the close interrelation of municipal functions is essential for good city government. The lack of coordination among municipal officials may be a reflection of some weakness on the part of the city administration, or it may be that the individual departments are more concerned about preserving their autonomy than in performing as part of a city government team. Some fire department administrative personnel have a tendency to believe that fire protection is so specialized that it has little relationship to other municipal government functions. The fire department is an important entity in municipal government, but to function most effectively it must cooperate with other municipal agencies for the mutual goal of improved public service.

CHARTER RESPONSIBILITIES

The charters of most cities include fire protection as a responsibility of the municipality. The provisions of some municipal and fire department charters may make specific mention of fire prevention, although they generally encompass overall fire protection, which may easily be interpreted to include fire prevention. An example of the former is the charter of the College Park, Maryland, Fire Department, which states the following purpose for existence: "Its object shall be the prevention and extinguishment of fires and the protection of life and property in College Park and vicinity."

Even if fire is not specifically mentioned, the responsibility for public safety is clearly drawn in the charters of practically all cities. There is no question that the public safety includes protection from fire, so it may be assumed that broad public safety provisions give charter backing for appropriate actions and measures in fire prevention.

Recent years have seen the role of the fire service expanded to include new responsibilities far beyond its original goals. As an example, the stated goals of the Fullerton, California, Fire Department are to provide services designed to protect lives and property of the people in the city of Fullerton from the adverse effects of fires, sudden medical emergencies, or exposure to dangerous conditions created by either man or nature.[2]

◆ FIRE SERVICE ADMINISTRATION IN MUNICIPAL GOVERNMENT

With thousands of incorporated cities and municipalities in the United States, it is understandable that there is a wide diversity of public fire services, ranging from volunteer or call groups in smaller communities, combination volunteer and career forces, to fire departments with specialized divisions for fire prevention. In the traditional municipal fire department, the fire chief reports directly to the mayor or other top administrator, an arrangement that makes fire safety responsibility equal to other municipal functions.

◆ FIRE PREVENTION FUNCTIONS

Prevention of fire is a primary goal of a fire department. As mentioned earlier in this chapter, fire prevention is considered an important component in the fire defenses of a community. Fire prevention refers primarily to measures directed toward avoiding the inception of fire. Among these measures are included fire and life safety education, fire prevention inspection, fire code enforcement, investigation of fires to determine causes, and investigation of suspicious fires. Analysis of information gained from the latter two functions can be of great value in evaluating and improving practices to help prevent fires.

There are many organizational approaches to assigning responsibility for fire prevention functions. Some fire departments do not consider fire prevention a full-time activity and may assign fire prevention responsibilities as collateral duties.

In several cities, the fire prevention bureau or fire marshal's office is a separate arm of government, with the fire prevention chief or fire marshal reporting directly to the city manager or chief executive. In some fire departments, fire prevention responsibilities are considered to be line functions for suppression forces. A number of federal and industrial fire departments are organized with fire prevention functions as a line responsibility for suppression personnel. These departments make sure that personnel are trained for such requirements. They are employed not only as firefighters but as fire "preventionists" as well.

The term *fire prevention bureau* is no longer the sole title for the arm of the fire department that carries out fire code enforcement, public education, and fire investigation duties. Fire loss management, risk management, environmental control, community risk management, community risk control, fire safety loss prevention, and a myriad of other titles may be found throughout the country. It is important to recognize that the word *fire* has a clear and distinct meaning to practically everyone, while some of these other titles may be baffling to the average citizen.

SARA Title III is a federal program that requires industry and related occupancies to maintain records on hazardous materials and their characteristics. Some fire departments assign monitoring duties related to SARA Title III plans to the fire prevention bureau. The fact that inspectors are checking on hazardous materials in the community makes this a logical arrangement. Although an actual chemical release may activate fire suppression forces, the routine oversight responsibilities can well be carried out by fire prevention personnel.

◆ THE FIRE PREVENTION BUREAU

In fire departments serving communities with populations of 25,000 or more, the most common organizational concept is a fire prevention bureau that functions as a separate arm of the fire department. Personnel are assigned to the bureau on a full-time basis and normally have no fire suppression responsibilities, except that in emergencies they may be pressed into service in fire fighting. In departments where there is a separate fire investigation bureau, fire prevention personnel may on occasion be called on to assist in investigations.

The fire prevention bureau coordinates all fire prevention activities within the fire department and makes important contributions to fire department public relations through its inspection and fire safety education programs. Fire prevention personnel are in direct contact with the citizens of the community.

In some fire departments, the attitude prevails that fire prevention work is for those who are physically unable to carry out active fire fighting duties. People who have been injured in the line of duty or who are suffering from some other physical disability that precludes their effective service in fire suppression are assigned to the fire safety bureau. This practice may be the result of a lack of understanding of the true role of fire prevention in the fire defenses of a community. Under the fire laws and regulations in effect in most jurisdictions, the fire department has the authority and obligation to carry out essential fire prevention measures. Furthermore, fire prevention work has some strenuous aspects. Conducting inspections, for example, may require climbing steps or other physical effort that may be beyond the capacity of the physically handicapped.

There is no doubt, however, that some excellent people have been assigned to fire prevention bureaus because of physical challenges that precluded their functioning in other sections of the department. In one example, the fire prevention chief was assigned to that duty because of a heart condition. Had it not been for this disability, he— a university graduate—would probably have become the chief of the fire department. As chief of the fire prevention bureau, he functioned in an outstanding manner and continued year after year in this capacity.

THE ONE-MEMBER BUREAU

The fire prevention bureau should be organized in such a manner as to permit clear lines of control. In a smaller community only one individual may be assigned to the bureau, which makes the matter of organization simple. The one-person fire prevention bureau can be quite effective. Arrangements should be made to allow this person to take care of duties in the field. Individuals can be assigned to fill in for the fire prevention office during absences from the office, vacations, or periods of illness.

Many smaller communities have outstanding fire prevention bureaus. Great care should be given to the selection of an individual to head a program of this type because of the range of duties required and because so much of the work involves contact with the public. After some years on the job and especially after becoming proficient at it, the lone fire prevention officer may become entrenched and possibly sidetracked from promotion in other areas of the fire department. Another problem in the one-person bureau is that the officer, because of constant public contact, may become better known in the community than the chief of the fire department, which could create internal problems. Close coordination and understanding between the fire prevention officer and the chief of the department are needed to maintain harmony.

GENERALISTS VERSUS SPECIALISTS IN THE FIRE PREVENTION BUREAU

In the fire department large enough to establish a fire prevention bureau consisting of more than one individual, it is important that the organizational concepts employed in the bureau be considered. There is a tendency to assign personnel to the bureau to get the job done, but without much thought to appropriate individual assignments.

In a smaller fire prevention bureau, consisting of from two to seven people, it seems most practical to assign all personnel as generalists. This would mean that any assignment for inspection, plan review, fire prevention education, or investigation could be handled by any member of the bureau. Assignments might be made on a geographic basis. Personnel assigned to one area of town are responsible for all inspections within that area, regardless of the type of occupancy. This arrangement has the advantage of providing inspections at the most reasonable cost, because of less travel involved. The inspector can go to the assigned section of the town, get out of the car, and go from door to door, conducting inspections for an extended period of time. This would be the case in the business district or in shopping centers located on the outskirts of the town. One disadvantage of this system is the possibility that having inspectors working different parts of town could bring about variations in inspection procedures or in interpretations of code requirements for similar facilities. A nursing home in the northwestern part of town, for example, might not be inspected in the same manner as one in the southeastern part of town.

On the other hand, some fire prevention bureaus, especially the larger ones, find it advantageous to develop and assign inspectors as specialists in a particular field. One inspector would specialize in flammable-liquids installations, for example, another in health facilities, while another might handle all industrial occupancies. If the inspection load in a community does not warrant full-time assignments in the inspector's area of expertise, the specialist may spend some time handling general assignment work. The disadvantages of the specialist system of inspections are that transportation costs are higher when inspections are not confined to one area and that inspectors have less opportunity to establish close contacts in the community than when working in one neighborhood. The advantages and disadvantages to the community and to the department should be thoroughly weighed to determine the best system for a particular organization.

Here, too, as with the one-person bureau, care must be taken to avoid dead-ending the specialist inspector because of expertise in a given field. Personnel assignments must take into consideration the promotional policies of the bureau and the fire department. Promotion to captain, for example, often requires several years' experience in fire suppression. The individual with a specialized background should not be penalized by not qualifying for movement within the department.

MAJOR CITIES' BUREAUS

A brief description of the structure of the Bureau of Fire Prevention of New York is included in order to describe the wide range of services necessary in a major city. Fire investigation and fire safety education functions are rendered by separate divisions in the department.

Bureau of Fire Prevention

The Bureau of Fire Prevention (BFP), staffed by 430 professionals, includes civilian fire protection inspectors, chemical and electrical engineers, supervisory and clerical workers and approximately 10 uniformed personnel working out of Fire Department Headquarters. The BFP generates approximately $33 million annually in inspection fees, permits and certificates.

The Bureau is commanded by the Chief of Fire Prevention, who reports directly to the Chief of Department, as mandated by the Administrative Code. That section of the code empowers the Bureau to perform the duties and exercises the powers of the Commissioner in relation to dangerous articles such as combustibles, chemicals, explosives, flammables, compounds, substances or mixtures. The Bureau also is responsible for the prevention of fires and the protection from panic, obstruction of aisles, passageways and means of egress, standees, fire protection and fire extinguishing appliances in theatres and places of public assembly.

The history of the Bureau dates back to 1865, when the Metropolitan Fire Department was formed from the Volunteer Department. Within the Metropolitan Department, there was a municipal department for the "Survey and Inspection of Buildings." The stated goal was the effective prevention of fires and the protection of life and property in the City of New York.…

The "modern" Bureau of Fire Prevention was formed on May 1, 1913, as a direct result of the Triangle Shirtwaist fire of 1911 and the Wagner Factory Investigation Commission. This marked the first recognition of the equal importance of fire prevention with fire suppression in the work of the Fire Department.[3]

SUITABLE QUARTERS

In the development and organization of a fire prevention bureau, it is important to provide suitable quarters for adequate operation and public contact. Unfortunately, some fire prevention bureaus are tucked away in a remote corner of the central fire station that is poorly located for public access. In some cities the central fire station is located in a neighborhood that has become undesirable from the standpoint of safety on the streets.

The location of the fire prevention bureau within the fire station may be a source of internal problems. In some cases it is necessary for the fire prevention bureau "customer" to go through sleeping quarters or day rooms occupied by fire fighting personnel to reach the fire prevention bureau. This is a bad situation, which not only may cause inconvenience for the public but also may result in resentment toward the fire prevention bureau on the part of the fire suppression personnel.

The physical location of the fire prevention bureau in the fire station or city hall may also be a factor in the public's attitude toward the fire prevention program. If the bureau is poorly located, perhaps in a building which is not in compliance with fire or building codes, the implication is that the bureau is of limited importance in the fire department and, in fact, in the overall municipal government. The fire prevention bureau should be located in a section of the fire station or city hall that can be easily reached by the public. Parking facilities should be made available for visitors.

COMMUNICATION WITH THE PUBLIC

The bureau offices should have adequate facilities for receiving telephone calls and electronic communications. Citizens calling to transact business with the fire prevention bureau should be able to reach a person who can supply the needed information without having to talk with several other people who may not be directly concerned with fire prevention. The fire prevention bureau should also have clearly established mailing and e-mail addresses.

Because of public uncertainty, the chief of the fire prevention bureau, or the fire marshal as some may be called, may receive communications, telephone calls, or visits that should properly have gone to the chief of the fire department. On the other hand, calls may be referred to the fire chief that should have been referred to the fire prevention bureau. With responsibilities for the overall administration and day-to-day operations of the fire department, the fire chief is not likely to be fully informed of routine fire prevention bureau activities. Perhaps some internal fire department method of screening calls and visits would help clear up confusions.

WORKING HOURS

Personnel in the fire prevention bureau are often assigned shorter working hours than other fire department employees. There are many exceptions to this rule, however, as fire suppression personnel are given shorter workweeks. At one time, assignment to the fire prevention bureau might have been considered desirable from the standpoint of working hours, but this view has all but disappeared with changes in working conditions for fire suppression personnel.

Practice dictates that the fire prevention personnel work on a schedule comparable to the headquarters office personnel and to the city government office forces. The workweek is normally 40 hours, although some cities have reduced this to 37 hours or fewer. Some fire prevention bureaus work four 10-hour days per week. In a number of cities, fire fighting personnel are now working 42, 48, or 56 hours a week. In some cities the workweek is down to 37 or 40 hours. Shifts are generally arranged on a 10-hour, 14-hour, or 24-hour schedule. In either case, employees have daytime hours off during the week, which allows them to participate in other activities during the normal daytime workweek. Such an arrangement cannot be made for an individual assigned to a 40-hour workweek in the fire prevention bureau. For this reason some cities are finding it difficult to encourage people to voluntarily enter fire prevention work. As a result, some cities are finding it necessary to upgrade fire prevention personnel and to take other steps to make the job more attractive.

Periodic reviews of time allocations are important in a fire prevention bureau. A typical analysis in a city of 100,000 (Gainesville, Florida) population appears here:

Summary: The Fire Rescue Department has reviewed the time allocation and inspection activity of a typical fire inspector based on one annual report. This report was evaluated and then reviewed by the fire inspector to reduce any anomalies. It was found that 59.5 percent or nearly two-thirds of an inspector's duty hours are utilized for conducting inspections. The remaining time was primarily utilized for certification maintenance, lunch, sick, vacation/holiday, administrative activities, and cross-utilization. This 59.5 percent was further evaluated to determine specific inspection activities. In this review, it was determined that 31.5 percent of time is spent on maintenance inspections of existing building stock (hazardous process, day care, citizen requests, assembly, all other existing systems/construction); and 28 percent of time is spent on other inspection-related activities (plans review, record keeping, burn permits, special events, complaints).[4]

Funding shortfalls have brought about some innovative reassessments of typical fire prevention bureau duty assignments. An example is cited here:

One city of over 75,000 population, Asheville, North Carolina, has divided its fire prevention bureau into two divisions, one of which handles only plan reviews, inspections, and tests relating to new construction while the other handles all existing structure inspections as well as fire investigations and complaints. Under this concept a deputy fire marshal is assigned to each of the three shifts. That person also coordinates inspections by fire suppression companies while the new construction personnel work

FIGURE **7.1** ◆ Fire inspectors spend countless hours studying fire and building codes. *(Columbia, South Carolina, Fire & Rescue Department)*

primarily day shifts so as to be available for contractors as needed. These working arrangements have served the department well and have reduced overtime pay demands.[5]

◆ SOURCES OF CONFLICT WITHIN THE FIRE DEPARTMENT

One of the major responsibilities facing the fire chief is that of bringing about a full understanding and appreciation of the fire prevention bureau on the part of other fire department personnel. Misunderstandings are not limited to those connected with working hours. Many other areas of misunderstanding arise in connection with the operation of the bureau. Friction may develop between bureau and other department personnel because of a feeling on the part of fire suppression personnel that the fire prevention staff are not exposed to the physical punishment associated with fire fighting and emergency medical services.

A conflict may also arise between the two segments because of the procedure followed in many communities of assigning automobiles to fire prevention bureau personnel. These vehicles are assigned for official use only, including investigation of overcrowding complaints and inspection of nightclubs. However, an elevated status may be attached to possession of a city car, especially if it may be taken home at night. There are many justifiable reasons for assigning automobiles to fire prevention bureau personnel. For example, it may be necessary for these individuals to respond to fires to carry out fire investigation duties.

Fire suppression personnel may also resent the fact that fire prevention bureau personnel are in the public eye more frequently. Inspectors are in daily contact with business owners, educators, operators of institutions, and other community leaders in the normal course of their duties. Fire suppression personnel have little public contact except on the fireground and on emergency medical calls.

◆ FIRE PREVENTION ADVISORY COMMITTEE

Another concept that has been found to be quite helpful in creating a better understanding of the role of fire safety is the fire prevention advisory committee, made up of members from all divisions of the fire department. One or two representatives might be selected from each of the battalions or districts within the department. The committee formulates the fire prevention program for the department and meets from time to time to evaluate their efforts and suggest changes in direction. Through the establishment of a fire prevention advisory committee, personnel throughout the department can be given an appreciation of fire prevention duties and responsibilities.

COMPANY FIRE MANAGEMENT AREAS

One of the most promising concepts is that of the fire management area program for utilizing company personnel for fire safety duties in their response territories. This concept was described by Springfield, Illinois, fire department officials at the time of implementation as follows: [6]

> Springfield has twelve fire stations. On the basis of still alarm districts, or at least approximately so, each station was given a geographical area, referred to as a Fire Management Area. This area was subdivided, assigning each shift one of the Sub-Fire Management Areas. Now we have 33 Sub-Fire Management Areas.
>
> The company is responsible for the following items in each area, by shift: a) hydrant inspections, once annually; b) target hazard identification and pre-fire planning to include area fire flow tests. Each Sub-Management Area team was required to identify the most serious life hazard, property hazard, and key hazard (hazard if property loss would affect the community the most—power plant, hospital, factory, etc.). Each company studied their sub-management area and picked the three targets. They then were assigned a pre-fire planning inspection to include the aforementioned area fire flow test. Using a "rule of thumb," they were then to calculate the required fire flow for the property. Material developed during these inspections is kept on file at each station and in our field operations office. Additionally, any fire code violations noted are forwarded to the Fire Safety Division for review. Referrals may also be made to Building, Housing, and Zoning Inspection Departments on referral forms. We now have over 300 targets identified and preplanned through the Fire Management Area system. A computer program for recall is in operation.
>
> Target hazards are videotaped and shown to all fire fighters and officers on a scheduled basis. Aerial maps are also used in pre-fire planning training.
>
> We continue to develop data and reinspect each year. In addition, fire losses are being tabulated by sub-management areas. This allows the individual company, and by shift, to understand more about their own fire problem, as well as the comparison to other areas of the city. This system has already produced several desired effects.
>
> Springfield attributed its success with this program to its participation in the master planning concept as described in Chapter 6. One of the values of the

concept was the recognition of Springfield as the first Class 1 city in the "rust belt" under the Fire Suppression Rating Schedule of the Insurance Services Office. Some cities have expanded the company fire management plan to include assignment of public fire safety education responsibilities to fire companies.

◆ **INTENSIVE INSPECTION PROGRAM**

Long recognized as a leader in fire prevention, the Cincinnati, Ohio, Fire Department has maintained an outstanding inspection program for many years. The origin of this program is described in Chapter 3; however, procedures used are worthy of study and emulation by other cities. The Fire Department inspects all buildings, including homes in the city. The system is described by the Department as follows:

Block Inspection
The Cincinnati fire prevention program has undergone revisions to reflect the constraints on man hours available and fire company manpower levels in a modern urban fire department. Today's fire department has a four-person minimum manpower requirement for each company. This minimum seldom allows the luxury for an individual to be available for follow-up inspections and has resulted in the need to "block inspect" or "unit inspect" almost exclusively.

With the increased customer service requirements that are expected from today's fire service personnel, there is a constant need to train members to maintain these levels of professionalism. Units of the Cincinnati Fire Department are rotated to the various "in service" training sessions in emergency medicine, haz-mats, rapid intervention and a myriad of other subjects. As a result of the increase in time spent educating and maintaining education of members of the Fire Department, residential fire inspections are scheduled based on requests from citizens. Categories of inspections other than one, two and three family residential units, are scheduled on a program as determined by the unit commander, subject to review by the district supervisor.

Several years ago, when a number of Cincinnati children perished in residential fires, an effort was made by members of the Fire Department to actually furnish and INSTALL residential, single station smoke alarms. It should be noted that there was resistance to this program from the City's legal department, but through perseverance, the program proceeded. The Fire Department solicited benefactors and installed the detectors via unit evening inspections known as "Safe Summer Nights." Nearly 5,000 smoke detectors a year for several years have been installed. The Fire Department feels that this accounts for the continuing decline in fire deaths within the city of Cincinnati.[7]

Tactical Inspection Unit
When the frequency of fires has increased critically in certain areas, a task group of fire inspectors in a concentrated inspection program is deployed to reduce hazards. These fire inspectors are detailed from various companies throughout the City under the command of a District Supervisor and concentrate on fire hazards within the critical area until they are rectified.

Cincinnati's fire prevention code enforcement efforts are directed toward serious violations where legal action is indicated. The Fire Department also is responsible for preventing environmental insult and has an environmental crimes unit to carry out these duties. The Fire Prevention Bureau of the Fire Division handles this work.

Cincinnati's change in emphasis from fire suppression and fire prevention activities to EMS responses is typical of a majority of fire departments nationwide. Information regarding the effect this change has had on the Department's long-standing fire prevention efforts is related here as an example of this phenomenon.

We no longer have the time to go door-to-door, unannounced and solicit the inspections. E.M.S. responses has a lot to do with that; in 1971 a "busy" fire company completed 1,000 responses a year and today a "busy" fire company completes almost 4,000 responses a year. Originally the goal of the Fire Department was to inspect ALL dwellings within the city each year and most businesses at least twice a year. As E.M.S. responses increased (today approximately 75% of our responses are for E.M.S.) it became obvious that we could no longer meet those earlier goals. New methods of setting goals were defined in which the local fire company commander sets the goals and one of the parameters for those goals was that one, two and three family residential inspections could be on an "as called for basis." Certainly these commanders can implement a yearly or biannual inspection program for troubled area where an outbreak of fires occurs and, in fact, we have done that several times in the last few years.

But what we found was the greatest intervention we could affect in residential fire loss was to make sure early fire warning was achieved. Enforcement

FIGURE 7.2 ◆ Many fire stations also serve as community centers enabling the imparting of fire prevention messages.

is still carried out but two things have affected our position on that. First, if you find a rental one, two or three family dwelling without smoke detectors you can cite the owner who will have to appear in court to face charges. However, you have left the home unprotected until you get the owner prosecuted and many times that is two or three months. We did not want to leave the residents unprotected; also, the prosecutor tells us that installing the detectors and THEN prosecuting will cause the judges to consider it very petty. (In other words: You have brought this man to my busy court for $10 worth of smoke detectors that you have already supplied?) The other influencing factor was that the largest supplier of our smoke detectors (a local nonprofit organization) has asked us not to prosecute because, I believe, some of its benefactors may be the targets of this prosecution and they will withdraw support. We no longer provide the random, unannounced, mandatory home inspections of one, two and three family dwellings. These we perform on a request basis only. However, we have a *very* concentrated smoke detector INSTALLATION program.[8]

◆ **FEES FOR SERVICES**

Some cities collect permit fees for certain fire prevention services. The rationale for the advent of this concept is explained in the following memo from the Western Fire Chiefs Association, Inc.:

Fire departments have historically been emergency services organizations and have given their services to the public free of charge. In recent years governmental leaders have questioned the high cost of providing fire protection and have looked for ways to offset these costs.

On the other hand, building departments for years charged various fees for building permits, plan review, and inspections. In many communities the building departments actually turn a profit and generate more revenue than it costs to operate their departments. The public and private sectors accept these fees as part of the cost of doing business.

Fire departments have been involved with plan reviews and building inspections where the building department was collecting fees and the fire department performed similar services at no charge. This has led to the conclusion by many fire departments that they should share in plan review and inspection fees. Also individuals or firms that engage in activities that require permits under the code should be required to pay for the special services required for the issuance of the permit.

In some communities the fee permit concept has been successful because the business community was made aware of the program and understood the intent from the time of implementation. In other cities the program had a very negative effect because the business people were uninformed and viewed it as just another unnecessary method of taxation.[9]

Although an increasing number of jurisdictions are collecting fees for their fire safety services, including inspections, plan reviews, and permits for certain occupancies and processes, few are covering the entire costs of their operations by this revenue stream. The fire administrator is still in the position of needing to justify expenditures for fire safety services. In fact, spending limitations imposed by state initiatives have forced several municipalities to look to the private sector for financial support to continue certain fire prevention activities, including public fire safety education programs. This may place the administrator in an uncomfortable position when private sector benefactors seek relief from fire code edicts, perhaps initiated by the same department employee who is using private sector funds to carry out fire safety programs.

◆ PERSONNEL TRENDS

A number of cities are using direct entry personnel for fire prevention duties. Dallas, Texas, for example, recruits people with at least two years of college directly into the fire prevention bureau. Their career ladder is limited to the fire prevention bureau supervisory positions, however. These individuals receive 710 hours of specialized training in subjects related to fire prevention.

Many fire departments are realizing the tremendous potential for women in fire prevention work. In some cases women are being assigned to fire prevention duties after completion of basic fire department recruit training. These individuals are members of the uniformed fire service. In other communities, women have been employed as civilian fire inspectors or fire prevention education specialists.

The use of women for fire prevention duties is not new. The 1923 Convention of the Oklahoma State Firemen's Association in Oklahoma City was addressed by Olga Juniger of the Texas State Fire Marshal's Office on the subject "Fire Prevention Work in Women's Clubs."[10] She was assigned full-time to fire prevention duties in her home state.

◆ STATE RESPONSIBILITIES

State agencies may exercise control over local fire code enforcement activities. As an example, Wisconsin law provides for local enforcement of the state's fire codes; however, the state provides funding through an insurance premium tax for the local jurisdictions to perform the necessary inspections and other related activities. The state Department of Commerce Fire Prevention Section employs several inspectors who verify that the local municipalities are actually conducting required inspections. Funding can be terminated if inaction is found.

◆ THE RELATION OF FIRE PREVENTION TO OTHER MUNICIPAL AGENCIES

Certainly, the fire department is the major municipal agency with an interest in fire prevention, and it is the one responsible for the overall control and enforcement of fire safety measures. As discussed briefly in Chapter 10, a number of other municipal agencies have a corollary interest in fire prevention as part of their major functions. To complete the picture of the place of fire prevention in municipal government, it is worthwhile to discuss the ways in which the entire municipal government is involved.

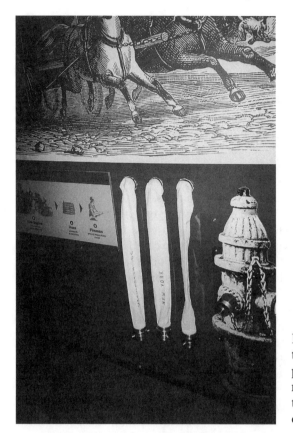

FIGURE 7.3 ◆ This Maryland Historical Society display illustrates problems out-of-town fire departments had with dissimilar hose threads during the Baltimore fire of 1904.

The municipal water department has a clear-cut tie-in to fire protection. Any improvements in the municipal water supply and distribution system will result in improvement in the overall fire protection in the community. For this reason the fire chief or fire marshal and the director of the water department should work closely together. Plans for expansion of the water system should always include fire protection considerations. As an example, a six-inch main might be adequate for domestic service in a given section of town, although an eight-inch or ten-inch main may be far more desirable from a fire protection standpoint. If the fire department cooperates with the water department, fire protection needs can be taken into consideration at the time of original development. Plans should also be coordinated if the water system is privately owned.

Another area of cooperation between the fire department and the water department is in testing of hydrants. Coupled with this activity may be the maintenance of fire hydrants, including painting and color coding.

Coordination with the police department is essential for effective fire prevention. In some cities there has been real or imagined friction between police and fire personnel, which may stem from rivalry or from misunderstanding of duties and responsibilities. Regardless of the reason, certainly, the public loses when misunderstandings mitigate services rendered. On the other hand, the public stands to benefit greatly through close cooperation between police and fire personnel in the interest of fire prevention.

In the course of their normal duties, police officers may observe conditions that should be reported to the fire prevention bureau for appropriate action. Overcrowding in places of assembly, for example, is more likely to come to the attention of the police than to the fire service. The police department can also be of great assistance in enforcing regulations covering the transportation of hazardous materials. Police officers are in a better position than the fire department to check on vehicles transporting dangerous materials to ensure the use of required markings, the use of designated parking facilities, and the observance of other safeguards. On the fireground, police and fire suppression forces traditionally work well together. The same kind of cooperation in other areas can advance the cause of fire prevention in the community.

The ultimate coordination of police and fire services is the combination police–fire department. Regardless of the pros and cons of the overall plan, it may safely be said that it is a mistake to expect personnel who are looked on primarily as police officers to carry out fire prevention responsibilities. The fire service has depended on a low-key public image to help in implementing fire prevention programs. The psychological impact of a person uniformed and armed as a police officer may create the opposite effect. One of the major advantages of the combined police–fire program is said to be the reduction of idle time. It can be argued that a fire department that is carrying out a wide variety of fire prevention and emergency medical programs in a busy organization has little time for police duties.

The local health department will probably have some responsibility for the inspection and approval of facilities from a sanitation standpoint. In the field of restaurant inspection, there is room for a great deal of cooperation between health and fire department personnel. The same may be true in inspection of day-care centers, boarding houses, hotels, and numerous other facilities where health department inspection and approval are required.

Coordination with the building department is in review of plans and specifications and cooperation in inspection for occupancy of structures within the community. Such coordination is extremely important. It should begin with coordinated reviews of preliminary plans and follow through to include cooperative inspections of occupancies. There are many examples of close cooperation between the two agencies. Traditionally, there have been misunderstandings between the two departments (building and fire) in some communities. Such misunderstandings and conflicts can result in compromising public safety—with dire results.

Fire department responsibility for building code enforcement was addressed in *Municipal Fire Administration,* a publication of the International City/County Management Association. This text states, "As a matter of strict logic, since most features of building regulation are for fire safety, the building code and fire prevention code could both be made a fire department responsibility. This is the case in German cities where one of the chief officers of the fire department is the city building official."[11]

Building codes, however, contain a number of requirements not directly related to fire safety. A fire chief who assumes responsibility for enforcing the building code must have adequate knowledge to carry out the duties. In some cities the building official's duties are assigned to the fire department.

Some communities, counties, and states have found it desirable to place all responsibility for enforcement of the fire-related sections of the building code in the hands of the fire department or fire marshal's office, with the building department retaining full responsibility for enforcement of all other sections. This arrangement seems to have worked well. An obligation is imposed on the fire department or fire marshal's office to provide adequate service with competent employees.

In some cities, the fire prevention code is enforced by the building department. All responsibilities relating to building inspection are thus concentrated in one agency.

This procedure has also been abandoned in several cities as the proficiency of fire services in this activity increases.

There should be coordination with the housing enforcement division of the municipal government. Housing codes cover many features of fire prevention, so coordination is mutually beneficial. The desirability of joint inspection may be reviewed on a local basis; however, in single-family occupancies, it would not be legal to conduct joint inspections without permission. The fire department may inspect such facilities on a nonobligatory, voluntary basis, although the housing section inspections are obligatory. It would not be in the best interest of either agency to have fire personnel enter on the strength of the housing inspector's right of entry. Both programs might suffer under such an arrangement.

Planning and zoning functions have considerable bearing on fire safety in any community. There should be no hesitation in coordinating fire prevention activities with those of planning and zoning to ensure recognition of necessary storage locations for hazardous materials and to ensure adequate clearances between buildings in the interest of preventing conflagrations. In some communities, the fire prevention chief or fire marshal actually sits as an ex officio member of the planning and zoning board.

The city purchasing bureau, which is generally responsible for the purchase of all supplies for the municipal government, offers an avenue through which the fire prevention bureau can work to have certain fire safety requirements established in city purchasing policies. For example, curtains and draperies and other materials used in public buildings can be specified to meet certain flame-retardant requirements. The purchasing agencies are usually willing to comply with such requests.

The sanitation department has a role in fire prevention. Disposal of trash and other combustible waste is, of course, a part of the fire problem in any city. The sanitation department can minimize hazards through uniform pickup procedures. The fire prevention bureau and the sanitation department should coordinate their efforts to ensure fire safety during extreme conditions. An example is a special trash collection during times of civil unrest to reduce the possibility of loaded trash cans being set on fire by disorderly persons.

Public utilities may be provided by private or public agencies. Regardless of ownership, fire prevention personnel should deal with these agencies just as they would deal with a municipal department. Areas of coordination include fire code enforcement, cooperation in responding to alarms, and other functions in which mutual interest is indicated. Cooperation with electric power, telephone, cable television, and gas companies is especially important.

The public school system is a government agency with great potential for cooperation in fire prevention endeavors. Cooperation may extend to arranging for fire department staging of fire safety demonstrations, classroom instructions, and cooperation with fire drills and inspections. This type of work has been most rewarding, but it can be carried out successfully only if there is complete understanding between fire prevention and education personnel.

The city library, although not thought of as an agency having a great deal of involvement in fire prevention, can be of value by providing space for displays and exhibits and by making available a wide variety of publications in the fire field. Books are available that may instill an interest in fire safety in the reader. Fire drill procedures in libraries are important too.

The judicial system has a bearing on fire safety. In most states the judicial system is primarily a state operation, although many states do have municipal courts, which are responsible for hearing cases involving housing, building, zoning, and fire code regulations. Thus, the fire prevention inspector may engage in enforcement activities involving cooperation with the courts and the city attorney.

Public welfare programs, which are generally carried out at the state level, offer some opportunities for coordination in fire prevention. A high percentage of public expenditures for fire protection, especially as related to responses to alarms of fire and emergency medical calls, goes to areas with low-income residents. This indicates a need for education in fire safety matters among welfare recipients. Very few jurisdictions have attempted to tie in fire prevention with welfare programs; however, avenues for development should be explored.

Cooperation on the fireground reflects on fire prevention and protection in several different ways. Fires in the community can have a tremendous impact on fire prevention efforts because they draw public attention to the subject of fire and demonstrate the need for fire prevention in a way that should not be underestimated. It is easy for people working in the fire safety field on a full-time basis to get the idea that everyone else thinks about fire prevention most of the time. This, of course, is not true, and many people give no thought to the subject until fire breaks out in their neighborhood.

Full cooperation between police and fire departments at the scene of a fire can go a long way toward smoothing out rough edges in public reactions. A motorist stopped by a hose line across a highway is usually irritated at the fire department, especially if the delay is lengthy. Under these conditions, the police department can probably do a better job of handling the public. By treating people in and around the fire scene courteously, they can preserve good public relations and allow the fire team to concentrate on their emergency operations.

Likewise, cooperation between the fire department and the water department is essential during major fires and can benefit both agencies. Deficiencies in water supply that may affect automatic sprinkler protection may show up during the suppression of a major fire. A fire emergency may also show the need for up-to-date information on water supplies, maps, and methods of designating hydrant locations and sizes, water main sizes, etc., as well as the need for augmenting water supplies during major fires and for providing additional hydrants in a given location. All of these factors can be used to press for public support for necessary water supply improvements. This will benefit the community in another way, too. Water supply facilities are recognized as a major factor in the classification of a municipality under insurance rating schedules.

The municipal streets or public works department can lend support to the fire department at the fireground. This support may include the provision of barricades at the scene of a major fire to assist in crowd control.

As important as the fire service and fire prevention functions are to the community, other municipal departments are equally involved with the safety and welfare of the public.

◆ SUMMARY

By tradition, fire prevention and fire protection in the United States have been primarily local government responsibilities. In the charters of most cities, public fire protection, which can easily be interpreted to include fire prevention, is named as a responsibility of the municipality. Court decisions have clearly indicated that the legal responsibility of a municipality to protect its citizens encompasses fire protection.

With thousands of incorporated cities and municipalities in the United States, the public fire services vary widely, from volunteer and call groups in the smaller communities; combination volunteer and career forces; to large, fully career fire departments with specialized divisions for fire prevention and training. In the traditional form of the municipal fire department, the fire chief reports directly to the mayor or other top administrator. In some municipalities, the fire department is placed under a director of public safety, who is also responsible for the police department. Another variation is the combination police–fire department, often known as a public safety department.

Fire prevention refers primarily to measures directed toward avoiding the inception of fire, among which are fire prevention education, fire safety inspection, fire code enforcement, investigation of fire to determine causes, and investigation of suspicious fires. In fire departments serving communities with populations of 25,000 or more, a fire prevention bureau generally coordinates all fire prevention activities. The bureau may range from a one-person operation to a relatively large operation with full-time personnel assigned to specialized functions, such as educational programs and technical inspections. Fire prevention personnel should be able to deal well with the public, and they should also have suitable quarters, easily accessible to the public, with adequate communication facilities to carry on their work.

Conflicts may arise between fire suppression personnel and the fire prevention bureau for several reasons: lack of understanding on the part of fire suppression personnel of the nature and importance of fire prevention work; differences in hours; social status associated with vehicles assigned to fire prevention personnel; and a general feeling that the fire prevention people have it easy and do not have to take the physical punishment that firefighters do. Some departments are requiring all fire service personnel to spend some time in fire prevention work; others make fire prevention training a part of their basic training; still others make fire prevention assignments a requisite of the rank of lieutenant or captain.

Other innovations to advance the cause of fire prevention are fire prevention advisory committees. These committees are made up of members from all divisions of the fire department—and sometimes also of people from outside the fire department.

Fire prevention is a concern of other municipal agencies, too, and the fire prevention bureau should cultivate opportunities to work with them in every possible way to get the message of fire prevention across to the whole community.

■ ■

Review Questions

1. Fire prevention and fire protection are primarily the responsibility of
   a. the federal government
   b. state government
   c. local government
   d. private enterprise

2. Responsibility for fire protection is usually established by
   a. charter
   b. state law
   c. county protocol
   d. the Constitution
   e. the fire department

3. The fire prevention bureau may also be called
   a. fire loss management
   b. risk management
   c. community risk control
   d. all of the above
   e. none of the above

4. A generalist in fire prevention could be assigned
   a. inspection
   b. plan review
   c. investigation
   d. fire prevention education
   e. all of the above

5. A source of conflict between suppression and prevention personnel is
   a. titles and status
   b. chain of command
   c. training opportunities
   d. assignment of automobiles
6. The department most likely to assist fire prevention with overcrowding in public assemblies is the
   a. building department
   b. police department
   c. zoning department
   d. planning department
   e. water department
7. Plan review is best coordinated with the
   a. building department
   b. police department
   c. engineering department
   d. planning department
   e. water department
8. Fire prevention refers to
   a. fire prevention education
   b. fire safety inspection
   c. fire code enforcement
   d. origin and cause investigations
   e. all of the above

Answers

1. c
2. a
3. d
4. e
5. d
6. b
7. a
8. e

Notes

1. Federal Emergency Management Agency, U.S. Fire Administration, *America at Risk* (Washington, D.C.: FEMA, U.S. Fire Administration, 2000), p. 15.
2. *Orange County Fireman* (Anaheim, Calif., Summer 2003), p. 10.
3. Fire Safety Education Fund Inc., *Fire Department City of New York* (Paducah, Ky: Turner Publishing Co., 2000), p. 58.
4. Gainesville, Florida, Fire-Rescue Department, Unpublished *Report on Fire Inspector Time Allotment* (2004).
5. Asheville, North Carolina, Fire Department entry to International Association of Fire Chiefs award program, 2003.
6. Personal communication to the author from J. P. Ward, Public Safety Director (ret.), Springfield, Illinois, 1998.
7. Personal communication to the author from Captain Bill Long, Fire Prevention Bureau, Cincinnati, Ohio, Fire Department, 2003.
8. Ibid.
9. Memorandum to Montana Vice President, Western Fire Chiefs Association from Uniform Fire Code Coordinator, 1986.
10. Oklahoma State Fire Fighters Association, *The First Forty Years* (Oklahoma City, Okla.), p. 89.
11. International City Management Association, *Municipal Fire Administration* (Washington, D.C., 1967), p. 232.

Instilling Positive Fire Reaction

8 CHAPTER

The reaction of the individual in a fire emergency is a critical element in life safety considerations, and the human behavior factor should be carefully analyzed in the development of fire safety codes. Public fire reaction is an integral part of fire prevention planning and education.

Any assessment of anticipated human reactions in the event of a fire must take into account individuals of all ages, health categories, physiques, occupations, and temperaments. The reactions of individuals to fires vary in relationship to these characteristics and to other factors such as the severity and location of the fire, the manner in which it started, the presence of other individuals within the fire area, and the specific condition of the individual at the time. The latter consideration includes factors such as whether the individual is awake or asleep, is intoxicated or under the influence of drugs, or is fully alert and capable. It also includes any other conditions that may have an effect on the individual's reaction to the specific situation with which he or she must cope.

◆ INDIVIDUAL DECISIONS WHEN FIRE OCCURS

Bernard M. Levin, Ph.D., formerly of the National Institute of Standards and Technology, has studied fire reaction for many years. He has concluded that people can usually be expected to make rational decisions during a fire; however, they may make inappropriate decisions that increase their danger because they do not understand the nature of fire and its growth and the elementary principles of fire protection.

Levin suggests that fire safety education programs address the phenomena of flashover, the speed with which a fire that seems small can grow to flashover, the concept of margin of safety, the size and nature of fires that building occupants can extinguish, and the value of closed doors and danger of opening doors. He explains the concept of "margin of safety" as the time between when everyone has left the area and when anyone in the room would be injured or killed.[1]

◆ FIRE REPORTING PROCEDURES

An example is the ongoing campaign to instill in the public an ability to properly summon the fire department. In addition to using stickers on telephones, some cities have resorted to billboard advertising and placing emergency telephone numbers on the sides of fire department vehicles in an effort to increase public awareness of fire reporting procedures. An effort is made to train individuals so that they will immediately report a fire. The almost universal use of 911 as a fire reporting number is helpful. The adoption of the enhanced 911 reporting system has also been very helpful in ascertaining correct addresses. Cellular phones are gaining an ability to define their locations to communications centers.

◆ HOME FIRE DRILLS

Another example of fire reaction training is the home fire drill program. This outstanding program, which has been under way for a number of years, was given considerable impetus by the Ohio State Fire Marshal's Office in the late 1950s. At the same time, a major project was undertaken by the Kansas City, Missouri, Fire Department, which resulted in widespread adoption in the greater Kansas City area. This program gained national stature and is being carried out in some communities as part of the activities of the annual Fire Prevention Week in October. Members of the public are encouraged to conduct a home fire drill at a predetermined time during Fire Prevention Week. In many communities, the program is known by the acronym EDITH: "Exit Drills in the Home."

The program encourages individuals to locate potential escape routes and to make plans to leave the dwelling without delay in the event of an emergency. This means determining all possible avenues of emergency egress and training family members to leave by those routes. The program also emphasizes the importance of leaving doors to occupied bedrooms closed to eliminate smoke travel as much as possible, although widespread use of corridor smoke alarms has raised a question as to the validity of this practice. Under the home fire drill program, individuals are taught to feel doors before attempting to open them. Participants are advised to check for smoke and, if it appears safe, to enter the corridor crawling on hands and knees, if necessary, to stay in the portion of the corridor that has the least smoke buildup.

A Milwaukee TV station conducted experiments to determine the ability of smoke alarms to awaken sleeping children. There were conducted as part of home fire drills. The tests revealed that children are not as responsive to smoke alarms as has been thought. They may fall back asleep only to awaken a second time. This factor needs to be considered in home fire drills.[2]

Another important part of the home fire drill program is a consideration of the need for all family members to assemble at a predetermined place once they are outside. This gives fire department personnel who arrive on the scene a means of learning which rooms might still be occupied to provide a better chance in any rescue operations. Fire department officers can immediately determine whether rescue operations are necessary. The officers will be able to check with a family member to determine where victims might be located.

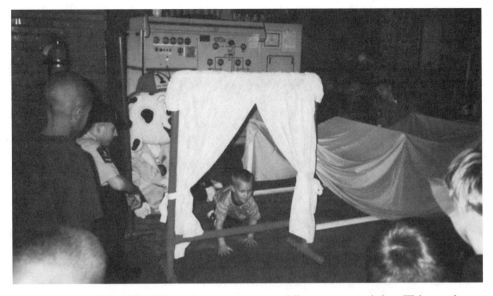

FIGURE 8.1 ◆ "Stay Low" is an important part of fire escape training. This mock-up is effective in achieving that goal. *(Lukfin, Texas, Fire Department)*

The home fire drill program emphasizes training in the need to call in an alarm without delay, although it recognizes that there are times when it might be more appropriate for the individual who discovers the fire to notify occupants and then notify the fire department.

Home fire drill training is a logical sequel to smoke alarm publicity. The tremendous popularity of these devices has resulted in the production of countless pamphlets, instruction sheets, and guides relating to this subject.

Closely related to both programs is training in procedures to follow when clothing catches on fire. People are taught the simple procedure of rolling over to control the flames. The procedure is known as "Stop, Drop, and Roll." These steps have saved many lives and minimized injuries in countless incidents.

The baby-sitter training program is an adjunct of home fire drill efforts to instill proper public fire reaction. This program includes other facets of safety; however, the fire prevention and fire reaction function is one of major importance. Training is given in measures to reduce the possibility of fire as well as in procedures to follow should a fire occur. Baby-sitters are instructed in means of evacuating infants and small children, in fire reporting procedures, and in other measures relating to fire reaction.

◆ SCHOOL FIRE DRILLS

Probably the most widespread program to condition public fire reaction is the fire drill program carried out in schools throughout the United States and Canada. Practically every state and province have requirements for fire drills to be held on a periodic basis in public, private, charter, and parochial schools. These drills are designed to ensure, so far as possible, the immediate evacuation of students from a school building

FIGURE 8.2 ◆ The Great Louisville Fire Drill draws large crowds to city parks each year. *(Capt. John Turner, Louisville, Fire & Rescue)*

in the event of a fire or other emergency. Most schools accept the fire drill program as a routine part of their operation and have shown little hesitation in actively requiring drills.

Many hidden benefits have come from the school fire drill program. Undoubtedly, individuals have reacted properly to fires in places other than schools as a result of training they received while they were in school. Fire drills in schools have probably helped individuals to more readily accept fire drill programs in institutions, industries, and other locations. Reactions to fires in theaters and other public assembly locations are also more positive as a result of conditioning received in school fire drill programs.

◆ INSTITUTIONAL FIRE DRILLS

In an institution, especially where nonambulatory individuals reside, reaction is usually expected only on the part of the institutional staff and visitors. The patients most often are not counted on for reaction of any kind, and institutional personnel initiate all evacuation measures. Complete evacuation, which is normally achieved in educational occupancy drill programs, is not generally expected in institutions. The drill may include horizontal evacuation to a place of safety within the structure rather than to a location outside the building. In some cases "defending in place" is recommended.

Institutions with ambulatory patients, including sections of mental, drug, and alcoholic rehabilitation institutions, may have programs in which each individual is part of the fire drill. Patients' participation in the fire drill will depend on their individual abilities.

Typical evacuation procedures are obviously not practiced in places of incarceration. In such institutions, measures are taken to eliminate as far as possible the chance of escape. This means that evacuation is at best most difficult and must be planned on an individual basis in each facility. Although inmates in a prison are most likely physically capable of departing from a burning structure, they must first be released from their cells by correctional officers.

◆ INDUSTRIAL FIRE DRILLS

Many years ago, little emphasis was given to fire drill programs in industry. Long hours coupled with arduous tasks left little time for fire drills. The 1911 Triangle Shirtwaist fire, as well as several other fires, including the Hamlet, North Carolina, poultry plant fire on September 3, 1991, have indicated the need for comprehensive fire reaction programs in this class of occupancy.

In arranging fire reaction programs for industry, officials must remember that shutdown time is expensive and that many machines and processes cannot safely operate unattended. Personnel can be familiarized with the location of fire exits, the location of fire alarm stations, and procedures to follow in the event of fire. As in the case of institutions, employees in industry may have duties to perform in reaction to fire that involve more than merely evacuating the structure. They may be responsible for obtaining and operating fire extinguishing equipment and calling the fire department. These duties should be described in an Emergency Action Plan. Illegal removal of products during fire drills is of concern in some industries.

◆ PROPER USE OF FIRE EXTINGUISHERS

Probably a major weakness in fire reaction programs is failing to adequately train people in the use of fire extinguishers and in their limitations. Records are replete with incidents of individuals attempting to control fires with extinguishers that were inadequate or improper for the conditions encountered. In cases where the individual also delays reporting the fire to the fire department while using the extinguisher, large losses may ensue.

A great deal of thought has been given to the relative role of the portable fire extinguisher in the fire protection field. This device depends entirely on proper human fire reaction for its effectiveness. Directions printed on the fire extinguisher are sometimes detailed and may be somewhat intimidating.

Programs for education in the use of fire extinguishers are widely available. The all-class fire extinguisher, which may be used on Class A, B, or C fires, has simplified training procedures considerably.

Traditionally, training in the use of fire extinguishers has been confined to individuals who have a direct responsibility for protecting specific locations, such as nurses in hospitals, custodians in schools, or maintenance employees in industrial plants. Even in industry, training of this type is not normally given to individuals who have little likelihood of using an extinguisher. Likewise, office workers in industry seldom are given such training. Fire extinguishers have extinguished thousands of fires in the incipient stage.

Figure 8.3 ◆ Fire extinguisher training is a vital part of industry's fire safety plan. *(Mead Westvaco, Luke, Maryland)*

◆ LARGE RESIDENTIAL OCCUPANCIES

Another place where fire reaction training could enhance fire safety is the multifamily residential occupancy. Apartment houses and condominiums with a number of units generally have interior fire alarm systems as well as fire extinguishers mounted at various locations in the structure. There are very few apartment houses or condominiums in which drills are carried out or in which fire extinguisher training is given.

Individuals living in apartments generally give little thought to the fire safety features of the structure. If a fire drill were staged periodically, tenants might feel a sense of security. For example, fire drills are fairly common in college dormitories. This procedure has saved lives on many occasions.

◆ PREVENTION AND CONTROL OF PANIC REACTIONS

A major consideration in educating the public for proper response is that of potential problems created by panic. In countless incidents, panic has played a major part in causing a large loss of life. Incidents include real and supposed fires as well as emergencies of other kinds that have occurred through the years.

In planning for public fire reaction, the planner must consider conditions that cause groups to panic, as well as the steps that may be taken to reduce the effects of panic. Panic is defined as "sudden, unreasoning, hysterical fear, often spreading quickly."[3]

Conditions that cause a group to panic include the existence of a real or imagined danger. Individuals in the midst of a panic usually are taken by surprise by the condition with which they are faced and may make a hasty evaluation of the possible dangers.

Psychological factors must also be considered in determining the possibility of panic occurring. People with feelings of insecurity, or an inclination to worry a great deal, or in a general state of tension may act unpredictably in a panic. These factors make it difficult to assess conditions that might create panic.

MEASURES TO COUNTERACT PANIC

A number of positive steps can be taken to reduce the possibility of panic. The importance of a few seconds in a panic cannot be overemphasized. On several occasions an individual has taken control of a panic by immediate action. For example, in a theater fire some years ago in Florence, South Carolina, a young usher went to the front of the theater, announced a fire drill, and asked everyone to leave in an orderly manner. Many people thought that this was in fact only a drill. They reacted calmly and departed the structure only to realize that there was a major fire in the building. Quick thinking on the part of an individual can be quite important, especially if the individual can be identified as having some authority.

Uniform as Symbol of Authority. The uniform has been found to be of considerable value in bringing about positive reaction to potential panic situations. In an explosion in Perth Amboy, New Jersey, a uniformed milkman was looked to for advice and guidance by many because his uniform represented authority and proper direction. Uniformed fire service personnel have been helpful in several other incidents.

Posting of Maximum Occupancy Signs. Posting maximum occupancies for structures may also help prevent panic. Posting should be based on measurable standards so that the number of people inside the structure at no time exceeds the exit capacities and predetermined limitations based on a square-footage formula.

Overcrowding is a very difficult condition to control, especially where people are moving in and out randomly. Of course, panic possibilities are not limited to places that are enclosed or under cover. Panic may occur even in an outside location.

It may be difficult for management to enforce fire capacity requirements in places where popular performances are being held. A fire official becomes unpopular in attempting to enforce requirements of this kind. This is especially true where the audience is excited as a result of the program being delayed because of failure of the feature attraction to arrive. It may be necessary to prohibit additional patrons from entering the facility. The facility manager should determine who must leave the premises if overcrowding must be relieved.

Emergency Lighting. Emergency lighting has also been found to be helpful in preventing panic. Many jurisdictions are requiring constant charge battery-operated emergency lighting units designed to give at least a minimum amount of light during a power failure. Exit lights are also important for preventing panic.

Exit Signs. Much study has been given to proper wording and color for exit signs. Some codes require green; others require red. The prevention of panic has been a factor in these deliberations and in the development of requirements. Some people associate red with fire and with exit lights. The other school of thought is that green indicates safety and that an individual seeing green will automatically go toward that door to reach a place of safety. There are valid arguments on both sides.

A combination exit light and emergency light is on the market. This device provides directed lighting under the fixture for exit visibility and, in addition, provides lighting of the exit sign.

Maintenance of Exit Facilities. An extremely important measure in panic prevention is maintenance of exits. No exit can be considered usable if it is necessary to obtain a key or remove obstructions. Individuals will react differently under the stress of even a small amount of smoke or heat. While it may be a minor problem to have to remove a chair that is blocking an exit, in time of stress this may be a monumental task. A chair or other blockage may cause a number of deaths because people cannot be expected to act rationally under such conditions.

Another example is the mistaken idea of some public assembly operators that certain exits may be locked when a less than capacity audience is anticipated. Their thought is that in the event of fire the individuals will see that the exits are chained shut and will automatically proceed to another exit just as they would change ticket window locations for quicker service. This thinking is fallacious because people are most likely to panic when they are forced into a locked exit, even though another exit is located nearby. This is especially true if real or supposed danger is immediately at hand.

Exits must be considered in the original construction of the building. Often, occupancy changes bring changes in exit requirements. Through periodic inspections, these changes can be noted by appropriate fire officials, and proper remedial steps can be taken.

Proper maintenance of exits means that all exits are properly marked and usable at times when their use might be anticipated. The exit lights must be on, the exit doors readily operable, and all readily accessible without blockage of any kind. There is a tendency to place ticket booths, tables, vending machines, and other objects in exit passageways without consideration of safety.

Public Address System. There should also be a means of obtaining immediate response from occupants to any emergency condition that might arise. Public address systems are extremely helpful. In some cases a public address system, operated by automatic control, has been used to replace the more normal fire bell or horn. Recordings over public address systems have been used to alert occupants to the existence of an emergency. Although a bell or horn is often used to signal an emergency, this sound does not always indicate an emergency condition to all occupants. Bells indicate telephone messages, changes of shifts, or the need for a supervisor, as well as other nonemergency conditions. Public assembly occupancies may be equipped with both fire alarm and public address systems. This is a most desirable situation. Emergency power is essential to successful use of public address systems for fire evacuations.

Advance Notice to Patrons. In an effort to prepare people for the possibility of a fire or other emergency within the premises, some jurisdictions require advance notice to patrons to encourage them to look for exits within the structure. Jurisdictions may mandate the use of a printed program, printed sign, oral announcement, or instructions on a screen to the effect: "Notice—for your own safety, look for the exits. In case of emergency, walk, do not run, to that exit."

The best way to prevent panic and encourage rational and adaptive responses to fire situations is to provide occupants with more rather than less information about the incident. This is not only helpful in preventing panic, but also in discouraging neglect or apathy in response to fire alarms. If people do not receive information concerning the true nature or circumstances of alarm signals, whether they are actual, nuisance, or false, they are likely to regard all of them as false because they rarely experience a fire alarm signaling an actual fire.[4]

FAMOUS PANIC SITUATIONS

Panic is not confined to fire situations. In May 1883, for example, the Brooklyn Bridge, spanning the East River between Manhattan and Brooklyn in New York City, opened for the first time. It was considered a great engineering feat of its day; however, many citizens were skeptical and did not believe the bridge was structurally sound. They were under the impression that this was just too great a feat for humans to accomplish. A week after it opened, many people were still apprehensive of the continued stability of the bridge. At that time a woman, in attempting to negotiate the steps of the Manhattan approach, fell down and screamed. Her screams caused many other people to think that the bridge was collapsing. In the panic that ensued, 12 people were killed, and scores were injured. Some of those killed, thinking that the bridge was collapsing, jumped off the bridge into the East River. Actually there was no damage to the bridge whatsoever.[5]

In a fire-related panic situation, 70 boys and girls were killed in the Glen Cinema in Paisley, Scotland. The local History Library in Paisley recorded the incident as follows:

In the disaster of 31st December 1929, 70 children aged between 18 months and 13 years died. In a room near the Gilmour Street entrance of the cinema (formerly The Good Templar Hall), a film had caught fire, sending up a cloud of thick black smoke which carried along a corridor into the main hall. On seeing this and thinking the building was on fire, the children rushed in a panic towards the exit at the opposite end of the building alongside the River Cart, only to find the exit doors were locked to prevent entry without payment. On arriving the fire brigade soon dealt with the small blaze and discovered the horrific scene; in their wild rush towards the exit some of the children had been trampled on, leaving piles of dead and dying children.[6]

More recently, panic occurred on January 29, 1956, when fire struck the Arundel Park Hall in Brooklyn Park, Maryland. At the time of this fire, approximately 1,100 people were in the hall. The fire broke out at approximately 5 P.M., and many of the patrons attending this church oyster roast believed that there was little danger; in fact, the fire was taken as a joke by many occupants.

Dr. John L. Bryan, then head of the Department of Fire Protection Engineering, University of Maryland, conducted an extensive investigation of panic and other factors related to the incident. Twelve patrons were killed in the fire. In addition, 250 people were injured.[7]

A quote from Dr. Bryan's report discusses panic conditions:

As the heat, flames and choking smoke began enveloping the hall the patrons grew more panicky and began to push toward several exits. Suddenly the lights were extinguished plunging the hall into darkness leaving only the red glow of flames overhead with the heavy smoke over all. This is when panic prevailed and people fought for exits and many were knocked down or lost their footing and were trampled. Many of the persons who had fallen to the floor were picked up and carried from the building. Fallen chairs and overturned tables from which all glassware had been knocked added to the confusion and the injuries.[8]

Dr. Bryan's investigation revealed that trained individuals who happened to be in the hall, including police officers and fire service personnel, reacted very much the same as untrained individuals. He further found that many people did not leave by the same door that they had entered, although this is commonly thought to be normal procedure. At least 12 people were found to have safely exited from the building only to reenter.

Twenty-one died and 57 were injured in a Chicago nightclub on February 17, 2003. Panic ensued when security personnel used mace to quell a disturbance.

◆ GUARD OR WATCH DUTY RESPONSIBILITIES

Within a given institution, mercantile establishment, or industry, security personnel may play a major part in the fire procedure program. The security officer is the only person on the premises during a majority of the hours in many such occupancies. The security officer may therefore have the responsibility for discovering the fire, reporting it, and carrying out fire control duties. This individual therefore may constitute a pivotal link in the entire fire safety operation of the facility. Unfortunately, this job is not always held by people who are adequately trained or capable of reacting decisively to emergency conditions. The security officer may not even be fully familiar with the entire facility.

REPORTING A FIRE

The importance of calling for assistance cannot be overemphasized. The security officer must be advised to call without delay and to take no chances of being overcome by smoke while trying to control a fire. There may be a tendency on the part of some security officers to play "hero." This may even lead them to start fires for publicity purposes. Individuals employed as security officers should be screened thoroughly, and a police record check should be made in every case.

◆ PRIVATE FIRE BRIGADES

Another means of ensuring proper fire reaction is through the organization of private fire brigades. Fire brigades are especially useful in industry and in institutions.

A fire brigade is organized to take immediate action in the event of fire. This action is of a first-aid nature and normally would only supplement, rather than take the place of, action by the public fire department. The public fire department may at times be delayed because of traffic conditions, inclement weather, bad roads, or simultaneous alarms. There are also occasions when the local public fire department is not capable of handling major fires in large industrial plants.

The private fire brigade has an advantage over the public fire service in handling fires in industrial plants, an advantage resulting from the brigade's knowledge of the plant. There should be brigade members in various sections of the plant or institution.

Fire brigade members have a primary responsibility to turn in an alarm on discovery of a fire. Except in a very large complex, this alarm will go to the local fire department as well as to other areas of the plant. In larger plants, it would go to the fire station located within the plant.

The plant fire brigade has a responsibility for operating fire extinguishers and hose lines. Brigade members must also understand the operation of fire sprinkler systems. Brigade members also serve as fire inspectors in their work areas and recognize problems inherent with excessive accumulations of trash and blocked fire equipment. They recognize that fire doors must be maintained in such condition as to permit closing in the event of fire.

As previously mentioned, some large plants and institutions maintain full-time fire departments. Their members generally spend a considerably greater percentage of their time on inspection and fire safety education duties than would be the case in a public fire department.

The size of a plant fire brigade will depend on the size and area of the plant as well as on the nature of processes carried out within it. As an example, a plant that handles flammable liquids in quantities will require more fire brigade members than would a plant handling sheet metal. Another factor to consider in determining the size of the fire brigade is the location of the plant. A rural plant will require more brigade members than would one in an urban area.

Another factor for consideration is coverage of all shifts. The provision of backup personnel for vacations and sick leave must also be considered. Because of limited training time available, people with previous training and experience in fire protection should be used when possible.

The individual selected as chief of the fire brigade should be from the supervisory ranks of the industry. In some plants, the fire brigade chief is selected by management, but in others the chief is elected by vote of the brigade members.

The fire brigade should include some members from the maintenance force of the plant. Pipe fitters, electricians, and other maintenance employees can be highly valuable to a plant fire brigade. OSHA provisions require training for these brigades.

FIRE BRIGADE DRILLS

Brigade drills should be conducted to ensure continued efficiency of operation. To be most effective, drills should be scheduled on a regular basis. Some drills are set up using repetitive procedures that have been practiced time and time again. It is most important to pose problems that require some thinking on the part of the participants. Drills held by industrial fire brigades, as well as those in institutions and mercantile establishments, should include occasional sessions in which public fire departments cooperate in prefire planning to give public firefighters a complete picture of plant fire protection features.

A fire bill or Emergency Action Plan operating procedure should be prominently posted to specify the duties and responsibilities of individuals in the plant or institution. The posting of such a plan is required under OSHA provisions where ten or more are employed. However, the fire brigade program should not function merely on paper.

As an added factor, there is a problem today with high stacking of combustible materials. There is also a trend toward windowless buildings, which makes the job of fire suppression more difficult. Another fire problem is the wide use of air-handling systems. These systems can be responsible for the rapid spread of smoke and fire throughout the premises, often in a manner that is not readily detected by fire suppression personnel.

Regulations of the U.S. Department of Labor's Occupational Safety and Health Administration (OSHA) relating to private fire brigades have discouraged some industries from establishing such units. OSHA requires training and proper protective equipment for individuals assigned to a brigade as part of their employment. Some industrial managers find these requirements so arduous that they decide to depend on public fire services rather than go to the trouble and expense of organizing a fire brigade.

◆ FIRE SAFETY MANUALS

In many occupancies, a formal fire brigade is impractical. In those locations, a fire safety manual may be used to promote proper fire reaction. These manuals form the basis of fire drill procedures and are tailored to the specific occupancy in which they will be used.

An example of a fire safety manual is the one in use in the residence halls at Oklahoma State University, Stillwater.[9] The manual contains information on fire exit drills, fire prevention and inspections, regulations for decorations, emergency lighting, trash chute fire problems, fire procedures, fire control, fire extinguishers, fire alarm systems, false alarms, fire bombings, fireworks, general safety, and first aid. A typical floor plan is included for use in developing fire evacuation plans.

◆ PUBLIC ELEVATOR FIRE SAFETY

Some jurisdictions require the posting of signs near passenger elevators to warn against using elevators in the event of an emergency. An example of this fire reaction program is the requirement that Maryland established in 1973, which requires a notice in all elevators: "WARNING: Elevator shall not be used in the event of fire. Use marked exit stairways."

The requirement was imposed as a result of several incidents in which individuals were killed or injured while attempting to use elevators when leaving burning buildings. The use of heat-sensitive door-opening devices has made the problem more acute because of the possibility that doors will open at the fire floor.

◆ FIRE REACTION SAFEGUARDS IN HOUSING FOR OLDER ADULTS

An outstanding example of guidance for instilling proper reaction to fire is the publication *Life Safety from Fire—a Guide for Housing for the Elderly,* a report prepared for the Architectural Standards Division, Federal Housing Administration of the U.S. Department of Housing and Urban Development.[10] The report is specifically designed to pinpoint problems in housing for older adults (age 65 and over). It was based on a study of thousands of fires, then reduced to a detailed study of 1,000 fires. It points out that many older adults are housed in buildings that were not specifically designed for them. It mentions problems in attempting to classify occupancy on the basis of known fire causes and in attempting to develop new classifications. New occupancy classifications are generally developed through experience with disastrous fires.

Few, if any, jurisdictions provide specific requirements for apartment houses occupied by older adults. The report suggests that a series of fires may need to occur before such requirements are imposed. The report emphasizes that older adults do not react in the same manner as younger people. A code requirement for 100-foot travel distances to exits is mentioned as being satisfactory for younger people who might be able to run that far under smoky conditions. However, this is an almost impossible feat for a person over 80. A 21-year-old subject could easily escape from a second-floor window; this would be very difficult for a 78-year-old person. An individual's reaction time is considered most critical. This is the time it takes the occupant to react to the fire and then to attempt to escape.

An interlocked chain of life safety operates in any occupied structure. Links of this chain are fire prevention, detection, alarm, escape or refuge, confinement, control, and extinguishment. All are necessary for complete protection. To ensure 100-percent life safety from fire is a theoretical ideal, and an effort must be made to make the links of the chain as strong as possible. Ignition temperatures and flame-spread characteristics of various materials are discussed in the HUD report. The flame spread of untreated red oak as the norm is rated at 100; cotton fabric in folds has a flame spread rating of 1,600 to 2,500. Fuel, smoke, and gases are other matters of concern in the overall fire picture when human reaction is considered. The fire resistance of materials in which the hourly rating is a factor is also of concern. A point is made of the matter of fire spread, which may be based on fuel available for burning. The presence of flame and the effect of smoke and heat on the human body are factors in finding probable paths of escape. The reduced oxygen content as a result of burning also must be considered.

Age, health status, and need for assistance in evacuation are important in connection with fire reaction, as are factors such as mobility, hearing, vision, and sense of smell. Approximately 35 percent of the ambulant population over the age of 65 have trouble climbing stairs and performing other complex movements such as climbing through windows. A study by the U.S. Public Health Service showed that two-thirds of all people 80 and over have significant hearing problems. This, of course, would have an effect on hearing the fire alarm, the fire itself, or shouted directions. Many older adults suffer from poor vision in dim light and have a tendency to be blinded by fire and flame. They also have a poorer sense of smell, which may have a bearing on original detection of a fire.

Another factor that might affect reaction is disorientation. Older adults sometimes tend to become confused and disoriented, which may cause them to be reluctant to use a given exit in the event of a fire.

Tolerance for heat, smoke, and gases also decreases with advanced age. This means that there is less likelihood of success in escaping from a fire. Shock may also be a factor and may be closely related to heart failure, which is an ever-present hazard of advanced age. Psychological factors affect the overall fire situation. For example, older adults typically collect objects and memorabilia in apartments and sometimes lose interest in keeping neat and orderly quarters. If a fire does break out, there is a possibility that the person detecting the fire will go directly outside and take no steps to sound an alarm or notify other individuals living in the structure.

An older adult may also be more inclined to attempt to put out the fire to maintain a sense of self-sufficiency. This attitude may be a matter of personal pride, and the person may feel that failure to control the fire may mean being forced to give up living independently and moving back with relatives or into an institution.

Many older adults view the contents of their rooms or apartments as their only significant material possessions. As a result, the decision to leave, especially if the fire cannot be immediately seen, may be an extremely reluctant one. In a number of instances, older adults have refused to leave their rooms during fires and other emergencies and have had to be forcibly removed by firefighters.

An older adult may easily overlook such simple protective measures as using wet blankets to keep out smoke from the corridor and pretesting doors for heat. Closely related to this problem is the tendency to follow familiar routes in an emergency escape, which is especially characteristic of older adults.

Reaction times for older adults are usually almost twice as long as those for young people. Specific designs of exit arrangements allow for decreased travel distance for older adults. In considering places of refuge, there is the possibility that individuals—especially older adults—may stay in their apartments, the safest place of refuge. Compartmentation has been shown to be helpful in minimizing the danger of fire spread from one apartment to another.

For fire safety, it is important to limit combustibles in hazardous areas and to provide automatic sprinkler protection. Fire drills can be effective, and they may even include response by fire department equipment on a nonemergency basis. Under no conditions, the report suggests, should the occupants be counted on to operate fire extinguishing devices, but emphasis should be placed on these individuals sounding alarms and evacuating the building.

The report just discussed addressed primarily older adults living in apartment buildings and other multiple family occupancies; however, over two-thirds of older adults (those over 65) live in their own homes. More than half of these live alone. The fire death rate for people over 60 is 20 percent higher than the national average, for those over 75 it is double the national average, and it is four times the national average among those over 85.

Persons over 65 represent the fastest growing segment of our population in the early part of the 21st century, and all indications are that this trend will continue. This factor makes reaching these citizens, whether they reside in a group environment or in their individual homes, a major goal of fire safety advocates. Often fire safety programs are geared to reaching those in group or apartment situations, giving little recognition to single-family dwellings occupied by older adults.[11]

◆ FIRE REACTION FOR SPECIAL OCCUPANCIES

Changing concepts of residential treatment create new challenges for those involved in developing proper fire reactions. As an example, several cities now have hotel-hospitals. This innovation enables people to move from a hospital, with its high daily room cost, to an adjacent hotel, which operated by the hospital.

Patients receive limited care in the hotel-hospital but can be taken back into the hospital without delay if necessary. This is especially useful for individuals who live some distance from the hospital or who have no one at home to provide posthospital care. Reaction to fire may change materially in the hotel-hospital because it usually has fewer staff than a hospital.

◆ FIRE REACTION IN HIGH-RISE STRUCTURES

High-rise buildings can present serious fire reaction problems. A number of factors may make it difficult for individuals to get out of these buildings in the event of fire.

Major problems develop because of the common use of the elevator for reaching all floors above ground floor. People occupying high-rise buildings seldom use stairways. This means that personnel employed in the building, as well as visitors, may not be familiar with the location of stairways. An example of this problem was encountered in a fire that occurred in a large southwestern city in a nine-story modern office building. All windows were of the fixed-sash type so that the balance of the air-conditioning system would not be disturbed, a common procedure in structures of this type.

The fire occurred during working hours; it was relatively minor and was confined to the basement of the structure. The problem was that smoke from the fire rapidly permeated the upper floors of the building through the air-conditioning system. The fire was brought under control, and fire fighting equipment was being prepared for return to service when a large executive chair came hurtling down from an upper floor. The chair hit the hose bed of one of the fire pumpers.

Personnel in the street looked up, only to see that occupants had gathered around a window that had been broken open by throwing the chair through it. The window opening was used to call for help. The aerial ladder, which was still on the scene, was raised to the floor level, and several people were rescued by ladder. Fire department personnel went up the stairways to the top floor and assisted other people to safety. The group of people assembled on the top floor were almost in a state of panic and had not given thought to using either of the stairways that were readily accessible nearby. For some reason, no one thought of trying any means of egress other than the elevator, which was customarily used.

Countless high-rise buildings throughout North America present similar evacuation problems. Normal fire reactions must often be modified in such structures.

Many studies have been carried out to develop improved fire safety procedures for high-rise structures. One of the most promising developments was the voice communication system for advising people about fire conditions. It has been thought that bell or horn alarms were adequate; however, in many conditions the voice alarm would be far more desirable, at least as a supplement to the fire alarm system.

Protection for high-rise structures should include automatic sprinkler protection. With this equipment, normal fire reactions might be entirely suitable because fires would be held to such minimal proportions as to enable egress by normal methods.

Lighting is another special factor for consideration in the high-rise occupancy. A power failure at a time of darkness can be extremely critical with or without a fire. The high-rise should be equipped with emergency lighting units or emergency generators to give occupants the opportunity to reach a place of safety without appreciable delay. They should not have to stumble down many levels in the dark to reach the outside of the building. This problem was reemphasized in the World Trade Center bombing in New York in 1993. A study of behavior during the explosion revealed some interesting findings. Over 100,000 people were evacuated in this incident in which six employees were killed and over 1,000 persons were injured.

The study verified that persons evacuating by stairway in high-rise buildings will continue their downward travel even as smoke conditions intensify. The need for evacuation training for all occupants, not just fire wardens, was amplified. Occupants need to have an understanding of smoke movement, the stack effect, and the dangers of falling glass to persons on the street below.[12]

Building operators in several cities have been given the option of permitting occupants to stay in place (except when their rooms are on fire) if the following conditions prevail: the building is of fire-resistive construction; fire procedure instruction cards are placed in each room; a public address system with emergency power is available to alert all occupants; each living unit is equipped with electrically connected smoke alarms; the building has full automatic sprinkler protection; at least one window is operable or breakable in each living unit; individual heating, ventilation, and air-conditioning units are provided, thereby limiting vertical and horizontal smoke spread; a complete fire alarm system is provided; one-hour fire separation is provided between living units and corridors; quarterly in-house fire inspections are conducted; at least two building employees are on duty at all times; and the fire alarm system is connected to the fire department.

An evacuation drill in a high-rise apartment building for older adults in a major city produced two cardiac arrests plus other disabling injuries. The city in which this occurred developed a modified or limited evacuation procedure in view of this experience. As a greater percentage of high-rise buildings are equipped with superior fire protection, the nonevacuation procedure will likely gain acceptance.

A study by Guylene Prouix of Canada's National Fire Laboratory confirmed that residents of conventional high-rise apartment buildings cannot be expected to evacuate immediately upon hearing a fire alarm. The study of a 1996 North York, Ontario, fire in which there were six fatalities included the following finding:

We found that occupants had long delays of two-to-three minutes before starting to evacuate in residential buildings where the fire alarm was activated. This delay is due to pre-evacuation actions, such as getting dressed, gathering valuables, and finding children or pets before starting to leave. Fire protection engineers are starting to use these findings in their designs.[13]

An example of a training program which addresses occupancy evacuation is the HOTEL EMPLOYEE LIFE-SAFETY PROGRAM (H.E.L.P.) taught by Las Vegas, Nevada's Fire Department in the hotels of that city. The program trains hotel security officers, management, and staff on what to do during an emergency prior to arrival of the fire department and how to assist firefighters after they arrive on the scene. It has proven to be quite successful.

◆ SUMMARY

An integral part of fire prevention planning and education is instilling awareness on the part of the public on how to act and what to do in the event of a fire, at home or in public places. Any assessment of human reactions that might be anticipated in a fire emergency must take into account individuals of all ages, health classifications, physiques, occupations, and temperaments. Fire severity, location, the manner in which it started, and the presence of the individuals in the fire area must also be considered.

One of the first things to stress is the importance of summoning the fire department promptly and giving the fire department dispatcher accurate information about the location of the fire. Stickers on telephones and on the sides of fire department vehicles—even billboards—can be used to keep the fire department telephone number constantly before the public.

Fire drill practices in homes, schools, institutions, and industry are an effective means of teaching people how to respond in a fire emergency in a planned, orderly way. Fire drill programs in schools throughout the United States have probably had a more lasting effect in conditioning the public to proper fire reaction than any other single effort.

There is a need for more widespread public education in the proper use of portable fire extinguishers and particularly in recognizing their limitations, both in the home and in public buildings. Although fire extinguishers have been used successfully on thousands of fires, many fires involving large losses have ensued because of failure to use fire extinguishers properly or delays in reporting a fire while attempting to use an extinguisher.

In planning measures for improved public fire reaction, the planner must consider conditions that cause panic and steps to take to counteract panic, especially in public places. Panic is easily triggered in a crowd, but certain physical measures can help to prevent it: enforcement of maximum occupancy limits, emergency lighting, clear designation and maintenance of exit facilities, use of public address systems for emergency instructions, and advance notice to patrons to look for exits nearest them to be prepared in case of fire. Closely allied to those measures is posting of signs near public elevators warning passengers not to use the elevators in the event of a fire and to use marked exit stairways instead.

Fire reaction safeguards in housing for older adults and in high-rise structures have received special emphasis as a result of several serious fires. Studies are being made of the special problems in both types of occupancies. As one report puts it, the essential links in the interlocked chain of life safety are measures for fire prevention, detection, alarm, confinement, control and extinguishment of the fire, and escape or refuge from the fire.

Not to be overlooked is the role of security personnel in fire protection of the premises in institutions, mercantile establishments, and industrial plants. Care should be taken to hire responsible people for this job and to give them adequate training to ensure proper fire reaction.

Private fire brigades, maintained by some institutions and industrial plants and trained in fire reaction procedures, have contributed immeasurably to public fire safety. Though intended primarily to supplement action by the public fire department, fire brigades can be critically important when the public fire department is delayed in answering the alarm or when it is overburdened during periods of civil disorder or natural disasters, or in areas where no public fire service is available.

■■■

Review Questions

1. In an assessment of anticipated human reactions in the event of a fire, one must consider
   a. height
   b. weight
   c. age
   d. sex
2. The condition of the individual at the time of the fire would include factors such as
   a. intoxication
   b. age
   c. occupation
   d. temperament
3. The ability of an individual to summon the fire department was improved by
   a. automatic sprinklers
   b. call boxes
   c. automatic alarms
   d. 911 reporting system
4. An experiment conducted by a Milwaukee TV station revealed that
   a. fire prevention does not work
   b. children may not respond to smoke alarms
   c. fire inspectors are not trained
   d. fire retardant clothes burn intensely

5. Institutional drills may include all of the following except
   a. complete evacuation
   b. horizontal evacuation
   c. staff actions
   d. inmate simulation
6. Sudden, unreasoning, hysterical fear, often spreading quickly is
   a. cowardice
   b. false bravado
   c. proper behavior in a fire
   d. panic
7. Panic can be reduced by
   a. emergency lighting
   b. exit signs
   c. public address systems
   d. maintenance of exits
   e. all of the above
8. Older adults are at risk from fire due to
   a. limited mobility
   b. reduced hearing
   c. poor vision
   d. lessened sense of smell
   e. all of the above

■■

Answers

1. c   5. a
2. a   6. d
3. d   7. e
4. b   8. e

■■

Notes

1. Bernard M. Levin, unpublished paper presented at National Fire Protection Association Fall Meeting, Nashville, Tenn., 1988.
2. "Smoke Detectors Need to Be More Alarming to Wake Kids," *Florida Times-Union* (August 7, 1993).
3. *Webster's New World Dictionary, Second College Edition* (Cleveland, Oh.: Simon & Schuster, Inc., 1995).
4. Personal communication from Mark Chubb, Southeastern Association of Fire Chiefs.
5. John L. Bryan, "Psychology of Panic," paper presented at Fire Department Instructors' Conference, Memphis, Tenn., February 18–19, 1958.
6. Personal communication from Ann E. Martindale, Local History Library, Central Library, Paisley, Scotland, 1978.
7. John L. Bryan, *A Study of the Survivors' Reports on the Panic in the Fire at the Arundel Park Hall in Brooklyn, Maryland, on January 29, 1956* (College Park: University of Maryland, 1956).
8. Ibid., p. 6.
9. *Fire Safety Manual,* Oklahoma State University Residence Halls (Stillwater: Department of Fire Safety, Oklahoma State University, 1997).
10. U.S. Department of Housing and Urban Development, *Life Safety from Fire—a Guide for Housing for the Elderly,* report prepared for the Architectural Standards Division, Federal Housing Administration, by Raymond D. Caravaty and David S. Haviland of the Center for Architectural Research, Rensselaer Polytechnic Institute, Troy, N.Y. (Washington, D.C.: U.S. Government Printing Office).
11. U.S. Fire Administration, *Fire Risks for the Older Adult* (Emmitsburg, Md., 1999), p. 3.
12. Rita F. Fahy, Guylene Prouix, "A Study of Human Behavior During World Trade Center Evacuation, *NFPA Journal,* National Fire Protection Association (March/April 1995), pp. 59–67.
13. *Education Section Newsletter,* National Fire Protection Association, Volume 18, Number 2, 1998.

Fire Prevention Efforts of the Private Sector

9 CHAPTER

Visitors from foreign countries who are in the United States studying fire protection activities are often amazed at the important role played by nonpublic fire safety agencies. In many countries, the private sector is of little significance in the development and implementation of fire protection procedures. By tradition, fire prevention activities in the United States have been motivated to a great extent by private organizations.

Fire safety activities of the private sector may be divided into three categories: those in which fire prevention is the primary function of the nonpublic organization; those in which fire prevention is a secondary effect of activities in the promotion of trade; and those in which fire prevention activities are carried out for the internal protection and well-being of the organization.

◆ ORGANIZATIONS WITH PRIMARY FIRE PREVENTION FUNCTIONS

NATIONAL FIRE PROTECTION ASSOCIATION

In the category of organizations having the prevention of fire as a primary responsibility, there would be little doubt that the National Fire Protection Association (NFPA), headquartered in Quincy, Massachusetts, would hold the top position. The objectives of the NFPA, as stated in its articles of association, are "to promote the science and improve the methods of fire protection and prevention; to obtain and circulate information on these subjects; and to secure the cooperation of its members and the public in establishing proper safeguards against loss of life and property by fire."

The Association was founded in 1896. It is a nonprofit educational and technical organization devoted entirely to preventing loss of life and property by fire and other hazards. Membership is open to any individual, company, or organization having an interest in these objectives. Although most members are from the United States and Canada, the NFPA is recognized worldwide for its impact in the fire prevention field, and it has members from more than 100 countries.

Activities of the NFPA include the developing and publishing of technical standards, codes, recommended practices, and model ordinances. More than 300 such publications have been developed, prepared, and updated under the guidance of 223 NFPA technical committees. More than 5,500 individuals serve on these committees, all on a volunteer basis. These technical committees usually include representatives from industry, public fire services, equipment manufacturers, the insurance field, governmental agencies at all levels, and other qualified individuals. Committees are arranged to preclude their domination by any represented groups that may have special interests in the subject at hand.

Standards are adopted only after detailed study and vote by the committee. Built into the process of developing and adopting standards, Calls for Proposals to amend existing documents or to shape the content of new documents are published. These public proposals, together with the committee action on each proposal, as well as committee-generated proposals, are published in the Report on Proposals for public review and comment. Public comments, along with committee action on each comment, are then published in the Report on Comments. Only after this public review and comment cycle has been completed is the final committee report brought before the membership for action. Once adopted by the NFPA membership at either an annual or a fall meeting and issued by the Standards Council, standards are published and made available for voluntary adoption. Many of these codes and standards have great influence and are used as the basis for legislation and regulation at all levels of government.

The NFPA has for many years published educational materials in the fire prevention field. These include school curricula (K–12), pamphlets, posters, occupancy studies, fire records, films, videos, and a wide variety of materials suitable for use in local fire prevention promotional efforts. More than 800 publications are available from the Association. The NFPA also maintains the world's most comprehensive fire protection library at its headquarters.

The NFPA provides technical and other topical fire safety information in its membership magazine entitled *NFPA Journal,* which is published six times a year. Other periodical publications for members include *NFPA Update,* a bimonthly newsletter containing current fire protection information and standards-related articles.

Among the many books published by the NFPA, the one best known is undoubtedly the *Fire Protection Handbook.* The first edition appeared in 1896, and through the years it has become the bible of the fire protection field. Other handbooks include *Industrial Fire Hazards, Automatic Sprinkler Systems, National Electrical Code, Life Safety Code,* and *Flammable and Combustible Liquids Code.*

The NFPA is the leader in public education programs concerning fire protection and prevention. Risk Watch® is NFPA's injury prevention curriculum for grade school students. "Learn Not to Burn" is the theme and focus of the NFPA's comprehensive public fire safety education program, which includes a national media campaign, a fire safety curriculum for school children, and an extensive outreach program supported by a regional field network of NFPA representatives. Fire Prevention Week is sponsored annually by the NFPA as a major national event designed to promote fire safety awareness. Sparky®, the Fire Dog, is the trademark creation of the NFPA.

The NFPA's Engineering Services Division has several field service projects working in specific phases of fire protection, including electrical, life safety, flammable liquids, gases, and marine fire protection. These special field service groups provide guidance in the application of pertinent standards in specialized areas.

The Association annually publishes statistics on the fire problem in the United States and maintains one of the world's most extensive fire experience databases. Combining its annual survey of fire departments and Fire Incidence Data Organization (FIDO) file on large fires and others of major technical interest, the NFPA publishes two dozen major annual overview reports on the nation's fire problem. It also provides numerous special topical reports and responds to thousands of individual and organization requests for data.

The Fire Investigations Department conducts investigations of major fires of technical or educational interest. Important lessons learned from the fires investigated provide input to NFPA technical committees and are vital in understanding how fires originate and how similar fires can be avoided in the future.

In 1982, the NFPA formed the National Fire Protection Research Foundation with the specific goal of identifying fire problem areas and developing solutions to those problems. Foundation programs help meet the urgent need to begin or continue research into hazardous materials, improved safety equipment for firefighters, better and more realistic code enforcement, improved public education, and a safer environment.

The NFPA has more than 75,000 members. Many of them are affiliated with one of the 14 membership sections. These include the International Fire Marshals Association, the Industrial Fire Protection Section, the Fire Service Section, the Electrical Section, and the Railroad Section. Each section has elected officers and carries out functions relating to fire prevention within its sphere of interest. Business meetings of these groups are normally held during the NFPA annual and fall meetings.

In an effort to provide better service to members, the NFPA has branch offices in Washington, D.C., California, Colorado, Delaware, Florida, and Kentucky. The Washington office serves as a liaison with all federal agencies on many matters in the fire prevention field. With the ever-increasing interest of the federal government in fire protection, coordination between federal and private agencies has become imperative. These NFPA offices have been valuable in coordinating efforts and gaining recognition of the part played by nongovernmental interests in fire protection.

OTHER FIRE PREVENTION–ORIENTED ORGANIZATIONS

The Home Safety Council, established in 1993, is a national nonprofit organization exclusively committed to preventing unintentional home injuries. The Council focuses on fire and burn prevention, which are among the top causes of unintentional home injury and death. Council programs and sponsored activities that address fire safety include The Great Safety Adventure®, Fire Prevention Week, *Risk Watch*®, Home Safety Month, the Safety Education Hero Award, and the CodeRedRover.org Web site for kids.[1]

The International Association of Fire Chiefs has as a primary objective the encouragement of fire prevention activities by local fire departments. The association is composed primarily of fire chiefs and their deputies in the United States and Canada.

The International Association of Arson Investigators is an organization of professional fire investigators. Its primary objective is the prevention of fire through the suppression of arson. The association encourages adequate training to enable its membership to fulfill responsibilities in this important field.

The International Association of Fire Fighters, an affiliate of the AFL-CIO, represents a large percentage of career fire department personnel in the United States and Canada. Local members have been active in fire prevention and have on occasion directly sponsored fire prevention displays at public gatherings.

The Congressional Fire Services Institute represents fire protection interests at the federal legislative level. The Institute provides liaison between the Congressional Fire Caucus and the fire protection community. Members of the Senate and House of Representatives belong to this caucus.

The Institute has a National Advisory Committee which reviews legislation relating to fire safety. Its work has been quite successful in coordinating fire service legislative initiatives.

The National Safety Council, an organization having major impact in the field of safety, includes fire prevention as an overall safety objective. The annual National Safety Congress is probably the largest gathering in the world of people with an interest in safety. Many of those active in the Council have fire prevention responsibilities as part of their overall safety duties. The Canada Safety Council has a similar role north of the border.

The Society of Fire Protection Engineers serves the professional field of fire protection engineering. The development and encouragement of fire prevention are major activities of the society.

The International Association of Black Professional Fire Fighters was formed to foster liaison among black firefighters. The organization has an interest in the promotion of fire prevention in high fire risk areas.

The International Fire Service Training Association, formed in 1943, develops and publishes training material for fire service personnel. Manuals on fire prevention subjects are included. Committee members from throughout the United States, Canada, and other countries gather annually under the auspices of Oklahoma State University to update the manuals and develop new material. A majority of states and provinces use these manuals for their fire service training programs.

The International Society of Fire Service Instructors was organized in 1960 by a group of state and provincial fire service instructors. The society has since expanded to include local training officers and other fire service personnel. The society has strived for improved instructor capabilities and has generally represented fire service training interests at the national level. Improved training in fire prevention is a goal of the organization.

The National Volunteer Fire Council represents the interests of the nation's volunteer fire service. Many volunteer fire departments have aggressive fire prevention programs.

The National Association of State Fire Marshals represents state fire marshals at the national level. This organization has actively supported fire safety education, smoke detector installation, and other abatement programs.

◆ ORGANIZATIONS WITH ALLIED INTERESTS IN FIRE PREVENTION

Private sector organizations having fire prevention as an adjunct of their operations may be divided into several categories. These include professional associations that issue standards having some bearing on fire protection, trade associations that deal in areas having some relationship to fire protection, organizations that develop model building codes, and the insurance industry.

PROFESSIONAL ASSOCIATIONS

Examples of the associations that issue standards having a bearing on fire protection include the Air Conditioning and Refrigeration Institute; the American Industrial Hygiene Association; the American Institute of Architects; the American Society for Testing and Materials; the American Society of Agricultural Engineers; the American Society of Heating, Refrigerating, and Air Conditioning Engineers, Inc.; the American Society of Mechanical Engineers; the American Society of Safety Engineers; the

American Water Works Association; the American Welding Society; the Institute of Electrical and Electronics Engineers; and the NFPA. All of these groups prepare standards and procedures that have some bearing on fire protection.

The American National Standards Institute does not prepare or write standards; rather, standards are prepared by some 250 separate standards developers, including those in the previous paragraph. Many of these standards are then submitted for approval as American National Standards (ANS). There are over 13,000 ANSI approved American National Standards. The American National Standards Institute is the coordinator and issuer of ANS designations.

TRADE ASSOCIATIONS

Trade associations that have some interests bearing on fire prevention include the American Fire Sprinkler Association, the American Forest and Paper Association, the American Hospital Association, the American Health Care Association, the American Gas Association, the American Iron and Steel Institute, the American Petroleum Institute, the Automatic Fire Alarm Association, the Compressed Gas Association, the Fire Equipment Manufacturers Association, the Gypsum Association, the Institute of Makers of Explosives, the Lightning Protection Institute, the Manufacturing Chemists Association, the Manufactured Housing Institute, the National Association of Fire Equipment Distributors, the National Association of Home Builders, the National Electrical Manufacturers Association, the National Fire Sprinkler Association, the National Propane Gas Association, and the Society of the Plastics Industry. These trade associations, as well as a number of others, have an interest in products and services that relate in some manner to fire safety. They often include fire prevention subjects in their publications and in programs presented for their membership.

To illustrate how a trade association may be concerned with fire prevention, the National Propane Gas Association will be reviewed in more detail. The industry represented by the Association markets a product that is inherently hazardous. To be marketable, the material must have the ability to burn—a characteristic that creates a problem in handling the material. To minimize dangers, the Association has taken steps to be sure that all individuals who handle the material are properly indoctrinated in safety measures.

The National Propane Gas Association, with headquarters in Chicago, has maintained active representation on the Liquefied Petroleum Gas Committee of the NFPA. Together they have participated in the development of standards for the safe use and handling of propane and have actively promoted employee training for member organizations. Liquefied petroleum gas associations at the state level are primarily concerned, as is the national organization, with the development of industrial and domestic uses of liquefied petroleum gas, but they also place considerable emphasis on safety programs of member companies. All these associations have actively worked to promulgate and enforce liquefied petroleum gas safety codes in many jurisdictions.

The American Gas Association, in addition to performing trade association services, operates testing laboratories in Cleveland and in Los Angeles for the purpose of testing all types of gas appliances. Safety testing is a major part of this function.

BUILDING CODE GROUPS

Building code development groups have a major impact on fire prevention. The building codes in effect in various jurisdictions throughout the United States and Canada play a major part in establishing fire protection requirements from a structural standpoint.

Building code groups that have operated at the national level in the United States include the Building Officials and Code Administrators, Inc., The International Conference of Building Officials, and the Southern Building Code Congress International. The three organizations have merged and now function as the International Code Council.

In addition, the National Conference of States on Building Codes and Standards works toward uniformity of building regulations. The group comprises state government representatives as well as other classes of members. Over 30 major cities participate in the Association of Major City and County Building Officials. Some of these cities develop their own building and fire codes.

INSURANCE INDUSTRY

As a result of the major fires that ravaged many American cities in the late 1800s and early 1900s, the property insurance industry played a major leadership role in the development and encouragement of fire prevention and fire suppression measures in the United States. Many advances came about as a result of the direct relationship between the fire insurance rate structure and the consequent implementation of local fire protection efforts.

In the 1980s, the insurance industry's emphasis on fire protection began to shift. The progression of insurance policies from covering only fire and lightning to package policies covering several property perils as well as liability lessened the insurance companies' centralized emphasis on the risk from fire alone.

In addition, strong competition developed among the insurance companies, which resulted in independent postures, lessening their joint efforts. The "mix" of business written by insurers has changed dramatically. Companies known as direct writers have concentrated their sales efforts on personal lines, whereas the large agency companies have emphasized commercial risks. Rather than supporting broad-based industry

FIGURE 9.1 ◆ Fire inspectors often train firefighters in operation of annunciator panels. (*Mead Westvaco, Luke, Maryland*)

efforts, more and more companies maintaining loss control departments specialize in the accounts each individual company wishes to insure.

There also have been major expense pressures on the insurance companies. As a result, many companies have reduced their contributions to trade organizations. That in turn has shifted those organizations' emphasis from public service activities, such as providing fire prevention activities and publications, to primarily dealing with federal and state government relations. Loss control services are aimed at specific occupancies. With the reduction in fire protection activities by the insurance industry, more and more of the functions they formerly performed are now conducted by governmental, educational, and private nonprofit organizations.

Those insurance organizations currently performing major activities in the fire protection field follow:

FM Global Group. This organization specializes in insuring major corporations' commercial property and is a leader in its market. From year to year FM Global Group ranks in the top five of those insuring large commercial property coverage based upon its direct premiums.

The FM Global Group is today's outgrowth of what was termed the Factory Mutual System. As is common in the insurance industry today, FM Global Group has evolved through mergers and acquisitions of other Factory Mutual System carriers whose operations were similar.

FM Global Group predominately insures only those large commercial property accounts considered as highly protected risks (HPR) or certain preferred property risks that do not meet the HPR standards but who meet the FM Global Group's rigorous underwriting requirements. FM Global Group is best known for its risk management and loss control services provided by its approximately 1,400 engineers.

FM Global maintains what has been termed the largest fire and natural hazards testing center in the world. It researches the effectiveness and reliability of industrial fire fighting equipment. It also tests electrical equipment and building materials. The issuance of the FM seal of approval signifies that the equipment bearing the seal meets FM Global Group's acceptability standards. It is the promulgation of its acceptability standards, the inspections of insured risks, and the research performed by FM Global that closely interrelates to the fire prevention and protection efforts of public and private authorities.

HSB Professional Loss Control. The purpose of this organization, which was owned by Employers Reinsurance Corporation and Hartford Steam Boiler and Inspection and Insurance Companies, is to insure well-protected industrial plants and institutions. It is now a subsidiary of the American International Group. The organization provides combined engineering and loss control survey services to insured factories, hospitals, and other large occupancies. The services of its nationally recognized fire safety laboratory are available for training of personnel in the maintenance and use of various types of fire protection equipment normally found in a major industrial plant.

INSURANCE SERVICES OFFICE, INC. (ISO, INC.)

This organization performs the functions that before 1971 were performed by the regional and single state fire insurance rating bureaus. These functions include

- classification surveying and assigning of public protection classifications to municipalities and fire protection districts
- promulgating loss costs (rather than rates) from which insurance companies themselves make rates

- the issuing of rules for determining program eligibility and classifications
- developing insurance contracts and endorsements
- gathering premium and loss information
- rating commercial buildings and their contents based upon the buildings' construction, occupancy, and fire protection features

In addition to its historical functions, ISO, Inc. performs similar services for insurance companies writing casualty lines such as liability, burglary, robbery, ocean marine, inland marine, and earthquake.

ISO, Inc. performs a wide variety of other services for the insurance industry including actuarial assistance and consultant services. It also combines risk factors and satellite imagery to pinpoint potential wildfire exposures, conducts on-site surveys of commercial and residential properties to determine their exposure to loss, and audits retrospective insurance policies to make certain the correct premiums are charged and paid. ISO, Inc. also provides claims information on bodily injury and property and vehicle claims from more than 256 million records, and it administers the Property Claim Service organization that reports and records severe weather and catastrophe information.

ISO, Inc. has recently developed a schedule, similar to its fire protection grading schedule, that grades municipal and county building code and enforcement measures for each community. This schedule is named the Building Code Effectiveness Grading Schedule.

In all states insurance policy contracts and endorsements, premium and loss reporting statistical plans, manual rules, protection classification schedules, and rates are subject to regulation by the state insurance department of that state. The statutory Insurance Code in each state specifically describes the basis upon which new and revised materials related to the foregoing will be considered as acceptable. In the United States the regulation of the business of insurance is the responsibility of the several states, and with very few exceptions, is not the responsibility of the federal government.

FAIR PLAN

Thirty-one states, the District of Columbia, and Puerto Rico have established FAIR plans for writing either basic property insurance or package policies. The acronym FAIR stands for Fair Access to Insurance Requirements, and the plans permit insurance coverage for properties that would otherwise find coverage difficult to obtain.

FAIR programs were developed primarily as a result of the urban riots in the late 1960s. In those states that have FAIR plans, all insurance companies licensed to do business in the state are required to participate. Each company is assessed to cover the plan's losses or share in the plan's profits based on the percentage of business the company transacts in the state as compared with the total premiums written. FAIR's inspections are conducted on individual risks, and coverage may be denied if the risk of loss is considered too severe.

UNDERWRITERS LABORATORIES INC.

Underwriters Laboratories Inc. (UL) is an independent not-for-profit organization having a comprehensive program of testing, listing, and reexamining equipment, devices, materials, and assemblies as they relate to fire protection and safety.

The laboratories, which were first established in 1894, test building materials of all types, fire protection equipment, gas and oil equipment, accident and burglary prevention equipment, as well as electrical appliances and equipment. The electrical testing program of UL, as an example, has had a major impact on fire protection, especially through the reduction of fires caused by the faulty design and manufacture of electrical devices and appurtenances. The UL label is internationally recognized as indicating that minimum standards for safety have been met by the product. Comprehensive tests have been devised to reduce the possibility of fire, shock, or injury of any type through the use of a product listed by UL.

An important feature of UL is its follow-up service program. UL's field representatives visit the factories where products are produced under UL service at least four times annually. The public is able to discern which items have been tested because of the use of a printed designation on the product. Over 17 billion labels are applied annually.

Underwriters Laboratories writes most of the standards used to investigate products. Those standards are regularly reviewed and revised on the basis of such considerations as field experience that indicates a product is unsatisfactory from a fire protection or safety standpoint. Product directories issued semiannually show names of manufacturers of products that are listed by UL.

The organization has established offices in many countries outside the United States in addition to its laboratories at Northbrook, Illinois; Santa Clara, California; Melville, New York; Camas, Washington; and Research Triangle Park, North Carolina. Underwriters Laboratories engineering subsidiaries operate in Hong Kong, Taiwan, Japan, and elsewhere. Inspection centers are maintained worldwide in more than 70 countries.

As an example of public utilization of a privately sponsored and developed activity, several jurisdictions require that all electrical equipment sold within the state or municipality be listed by UL or another recognized testing laboratory. Maine has had such a requirement for a number of years. This came about because this state, in the far northeastern corner of the United States, was becoming a dumping ground for inferior or faulty electrical equipment from all parts of the country. North Carolina enacted a similar requirement in 1934, and Maryland followed in 1973.

◆ PRIVATE ORGANIZATION EFFORTS IN FIRE PREVENTION

Fire prevention activities carried out by private organizations for the improvement and protection of their own properties are many and varied. In the majority of business establishments, industries, and privately operated institutions and schools, the primary concern with fire prevention is brought about by the local fire department, fire marshal's office, and insurance carriers. Some of the more progressive businesses maintain extensive internal fire prevention programs.

Some industries handle fire prevention at a local level with a person or persons being assigned in each plant for such duties. In other cases, a floating fire prevention person may service a number of plants out of a central location.

SELF-INSURANCE PROGRAMS

An example of a program in which there is a direct relationship between fire prevention efforts of private enterprise and expenditures is the self-insurance program. Under a self-insurance program, a business organization or government body accepts the risk connected with fire, either on a partial or complete basis. The industry or government body elects to handle fire losses with its own resources, rather than depending on commercially available fire insurance and/or comprehensive liability insurance.

A self-insurance program places a direct responsibility on management, whether the program is operated by a governmental agency or private organization, to be sure that all proper precautions are taken to avoid fire. Because there is a chance that losses might not be fully replaced or that any reserve fund might be depleted, an obligation exists to ensure minimal losses.

The most effective self-insurance plans are those in agencies or businesses that have a spread of risk arrangement so that the possibility of a total loss at one location is avoided. A state government, for example, has hundreds of fixed installations; the likelihood of a number of these installations having a major fire during a given year is highly remote. As long as the reserve to handle the anticipated losses can be maintained, the plan should function satisfactorily.

One unfortunate feature of the self-insurance plan is the possibility the reserve fund will be used for other purposes. If, for example, a company with a self-insurance program decided to open a new operation and start-up funding were required, there would be a real temptation to tap the fire-loss reserve for this purpose.

A self-insurance program must consider how to enforce the requirements necessary to maintain low fire losses. The usual procedure is to give someone or some group the responsibility to determine what measures must be taken to make the place reasonably safe from the standpoint of loss. In industry a consultant firm is often used to provide outside inspections for the company.

Controls needed to ensure continued satisfactory operation include plan-review and inspection functions and detailed analysis of all special hazards within the industry or institution. It is also necessary to have a soundly developed program for training maintenance personnel and others directly engaged in the day-to-day operation of the structures in question.

EDUCATIONAL INSTITUTIONS

Several private universities offer programs related to fire safety. The Worcester Polytechnic Institute, with its Center for Firesafety Studies in Worcester, Massachusetts, has carried out research in a number of fire safety matters and offers a master's and a doctorate in fire protection engineering.

FIRE DEPARTMENT ASSOCIATIONS

Although made up primarily of members of public agencies, regional, state, and local fire associations are actually private organizations. These associations exist in practically every state and province.

Many of these associations have both career and volunteer fire department membership. In some states, career and volunteer fire service groups have organized separate associations functioning within the same state. In either case, fire prevention may well be considered a major activity of the organization. Most states have fire chiefs associations, as well.

Generally, a fire prevention committee is one of the standing committees of the association. Its function is to encourage the promotion of fire prevention by the member departments. The committee may hold meetings for the purpose of distributing fire safety information, and in some states it conducts an annual fire prevention contest.

An example of a fire prevention group within a state fire association is the Fire Prevention Committee of the Maryland State Firemen's Association. This committee carries out an active program of fire prevention within the state and conducts several fire prevention contests each year. These contests include fire prevention activities by individuals as well as the fire companies within the state.

A number of county fire associations likewise sponsor active fire prevention programs. These programs may directly reach the public or may merely be designed to encourage fire prevention activities on the part of member companies.

LOCAL PRIVATE–PUBLIC COOPERATION

The Safety and Fire Education Committee (SAFE) that functions in Columbus, Georgia (population 186,000), is an excellent example of public–private area cooperation at the local level. Columbus has a combined city–county government and is home to a number of industries, a university, a major military base, and an insurance industry giant.

> The concept of SAFE is that through forming a partnership between fire-fighters, corporate and media professionals we could initiate, develop and implement new pro-active fire safety education programs and look for methods of enhancing programs already in place. Corporate personnel are familiar with setting goals, formulating workable plans and following through to a successful conclusion. Partnering with media personnel helps to get our message out in a positive manner to as many people as possible. The firefighters are a valuable resource for plan ideas, technical support and manpower for actual implementation of the programs. The concept is being very well received by all corporate and media personnel we have contacted.

> Current projects include the purchase of a combination sprinkler and fire safety house; an endowment to fund off-duty firefighters to staff the fire safety house; and

FIGURE 9.2 ◆ Columbus, Georgia, has a committee that includes fire service, corporate, and media professionals who initiate fire safety education programs. *(Columbus, Georgia, Fire Department)*

the introduction of a Fire Safety Olympics program in the elementary schools. Future goals include implementing a fall and fire prevention program for seniors and providing carbon monoxide detectors to low-income and older adult residents. SAFE also hopes to fund grants for fire prevention programs for small rural volunteer fire departments in the area.[2]

◆ SUMMARY

Fire prevention efforts in the United States have been motivated predominantly by private organizations, rather than by public agencies, as in most other countries.

Among organizations with a primary function in fire prevention, the unquestioned leader is the National Fire Protection Association, which has membership in over 100 countries. It is a nonprofit educational and technical association devoted to promoting scientific methods of fire protection and prevention and circulating information on these subjects to guard against loss of life and property by fire. Among the many publications of the NFPA, probably the best known is the *Fire Protection Handbook,* which has become the bible of the fire prevention field. Others include educational material and various periodicals, records of fires and fire loss statistics, and over 300 publications on technical standards, codes, recommended practices, and model ordinances, which are developed, prepared, and undated under the guidance of more than 200 technical committees.

Among other fire prevention–oriented organizations that have contributed greatly to the cause are the Home Safety Council, the International Association of Fire Chiefs, the International Association of Arson Investigators, the International Association of Fire Fighters, the Society of Fire Protection Engineers, and the National Safety Council.

Organizations with allied interests in fire prevention are noncommercial organizations, professional associations and institutes in special fields that prepare standards and procedures having some bearing on fire protection, and a long list of trade associations devoted to the promotion of trade in products and services that relate in some manner to fire prevention.

Building code development groups have made a major impact on fire prevention. Operating at the national level is the International Code Council.

The insurance industry has been a dominant influence in the development and encouragement of fire prevention measures in the United States, and there is no doubt that many improvements have come about as a result of their direct relationship to insurance rate structures. Here, too, the associations and mutual systems representing a number of insurance companies have made significant contributions.

Insurance Services Office, Inc. establishes insurance rates and is responsible for the grading of municipal fire defenses. In all states, insurance rates are subject to review by the state insurance commissioner.

Underwriters Laboratories, Inc. has long been an important force in fire prevention through its comprehensive testing programs for equipment, materials, and assemblies that relate to fire protection and safety. It has earned an international reputation for the UL label.

Some private business organizations maintain extensive internal fire prevention programs for their own protection. Others elect to operate under a self-insurance coverage. The federal government might be considered a leading self-insurer. A self-insurance program puts direct responsibility on management to be sure that all proper precautions are taken to avoid fire.

And last, the contributions of educational institutions that conduct courses in fire science and administration, and of firefighters' associations that are active in fire prevention education, must also be counted in the monumental effort that has been made by the private sector in this country toward the cause of fire prevention.

■■■

Review Questions

1. The National Fire Protection Association was founded in
   a. 1853
   b. 1873
   c. 1896
   d. 1925
   e. 1947
2. Activities of the NFPA include
   a. code enforcement
   b. developing technical standards
   c. mandating standard operating procedures
   d. establishing laws
3. Which is not an organization promoting fire prevention?
   a. International Association of Fire Chiefs
   b. International Association of Arson Investigators
   c. National Association of State Fire Marshals
   d. International Association of Fire Fighters
   e. International Municipal Fire Fighters

4. The American National Standards Institute
   a. prepares standards
   b. writes standards
   c. approves standards
   d. enforces standards
5. Those associations that have an interest in fire prevention include
   a. Central Station Alarm Association
   b. Compressed Gas Association
   c. National Propane Gas Association
   d. all of the above
   e. none of the above
6. The Insurance Services Office, Inc.
   a. sets insurance rates
   b. rates fire departments
   c. investigates fires
   d. both a and b
   e. none of the above

■■■

Answers

1. c
2. b
3. e

4. c
5. d
6. d

■■■

Notes

1. Communication to the author from Meri-K Appy, President, Home Safety Council, 2003.

2. Communication to the author from David L. Jones, Fire Marshal, Columbus, Georgia, Fire and Emergency Medical Services, 2003.

CHAPTER 10 ◆ Fire Prevention Responsibilities of the Public Sector

In the United States, governmental agencies have played a less prominent role in fire prevention than have their counterparts in other countries. However, the trend toward heavily urbanized areas has made greater governmental responsibility in the field of fire prevention necessary, and the influence of governmental agencies at all levels is steadily increasing.

In this chapter, the role of the federal government as well as those of state, provincial, and local governments will be outlined. Rapid changes are taking place in the sphere of governmental responsibilities, and activities by the federal government that were unheard of 30 years ago are commonplace today.

The major change has taken place at the federal level. Not too many years ago, federal responsibility was confined primarily to direct protection of federal properties and fire safety enforcement relating to a rather narrow group of activities involving interstate commerce. Current interpretations of the term *interstate commerce* have broadened the scope of federal responsibility and operation in that field. The findings and recommendations of the National Commission on Fire Prevention and Control, established by an act of Congress in 1968, are also having a strong influence on fire prevention responsibility in the public sector.[1]

Internal fire prevention efforts are broad in scope and encompass practically all federal agencies in some manner. A number of larger federal government installations maintain their own fire departments with personnel having responsibility for fire safety inspections, enforcement of fire prevention regulations, and public fire prevention education, in addition to their fire suppression duties.

Standards for fire safety within the federal government in the United States are generally set by individual agencies. Agencies maintain guidelines and regulations that must be adhered to by their employees. Regulations and standards of construction vary within the federal government, although the advent of the General Services Administration as a property management arm of government has probably brought about more uniformity.

Many federal agencies include fire prevention and evacuation training as a part of the orientation program for new employees. The responsibilities of certain federal agencies, especially the armed services, demand a high degree of attention to fire prevention because of the nature of the equipment they employ. The seagoing services, the U.S. Navy and U.S. Coast Guard, place a great deal of emphasis on fire prevention because of the grave problems related to fires aboard ship. The same condition prevails with aircraft assigned to the military services.

Although primarily aimed at enhancing fire prevention within federal facilities, government efforts in internal fire prevention have had a profound effect on the fire prevention behavior of the public. There is no question that fire safety training and actual fire experiences while on active military duty have made a lasting impression on significant segments of the present civilian population.

In a similar way, military reservists and National Guard members are exposed to fire prevention education through weekly or monthly drills that may include an occasional fire prevention lecture. Likewise, training programs offered to federal employees can be of value in home fire prevention and fire response conditioning.

Federal agencies with fire prevention responsibilities that have an impact on members of the general public include practically all cabinet departments and a number of independent agencies. Summaries of the fire prevention work being done by the various governmental departments and agencies follow.

◆ U.S. DEPARTMENT OF STATE

The Department of State has had several foreign aid programs relating to fire prevention as part of public safety efforts. These programs, which have operated under the Agency for International Development, have resulted in the imparting of recommendations and funding for the building of fire suppression and prevention forces in certain underprivileged areas of the world. Fire safety of U.S. embassies is also a responsibility of this department.

◆ U.S. DEPARTMENT OF THE TREASURY

The Internal Revenue Service, a Department of Treasury agency, engages in activities related to fire prevention that have an effect on members of the general public. Responsibilities of the Internal Revenue Service include the determination, assessment, and collection of internal revenue taxes. In performing these duties, agents of the agency may uncover attempts at fraud that included set fires. A number of fraud by arson cases have been thwarted by this agency.

◆ U.S. DEPARTMENT OF DEFENSE

The Department of Defense includes the Army, Navy, Air Force, and Marine Corps, which is part of the Navy. These elements are considered to be military departments within the Department of Defense. In addition, several other agencies are within the Department.

DEPARTMENT OF THE ARMY

An Army activity having a bearing on fire prevention and protection is the maintenance of explosives ordnance disposal units stationed throughout the country.

These highly trained units are available at the request of local fire or police personnel to remove and render harmless military explosive devices that may be found in a community.

Several research projects carried out by the Army have likewise contributed to the field of fire prevention. These include the development of fire-retardant materials.

DEPARTMENT OF THE NAVY

The Department of the Navy, which includes the Navy and Marine Corps, has made a substantial contribution to the fire prevention field through extensive research activities. Although much of the research relates to the development of fire suppression devices and equipment, some effort has been directed toward fire prevention. Fire suppression research is, in fact, an effort to bring about more effective control of fire, which contributes to fire prevention through reduction of fire spread.

DEPARTMENT OF THE AIR FORCE

The Department of the Air Force has devoted a great deal of research to the development of specialized equipment for fire suppression. Funding and technology have been available to permit extensive testing of new approaches in fire suppression before they are put to widespread use.

As with the Navy, a number of Air Force fire suppression advancements have had a residual effect in fire prevention. The very nature of fire suppression devices and procedures has the indirect effect of calling attention to the subject of fire.

◆ U.S. DEPARTMENT OF JUSTICE

The Department of Justice is the enforcement agency for the Americans with Disabilities Act. This act has a strong influence on building and fire codes. An agency of the Department of Justice, the Federal Bureau of Investigation pursues activities that contribute to the cause of fire prevention as does the Bureau of Alcohol, Tobacco and Firearms.

FEDERAL BUREAU OF INVESTIGATION

The Federal Bureau of Investigation has for many years provided laboratory examination services to local law enforcement agencies. Examinations conducted include those of materials found in and about a fire scene. Many individuals have been prosecuted for arson and unlawful burning based on examination of evidence by the Federal Bureau of Investigation Laboratory.

The Federal Bureau of Investigation has been of considerable assistance in arson investigation through the services of the National Crime Information Center and the extensive fingerprint identification program carried out by the agency. With authority over fugitive felons crossing state lines, the bureau has been of assistance in apprehending wanted arsonists.

BUREAU OF ALCOHOL, TOBACCO AND FIREARMS

The Bureau of Alcohol, Tobacco and Firearms of the Department of Justice enforces federal statutes relating to explosives. The agency regularly inspects and licenses explosives dealers, manufacturers, and users when any aspect of the operation may involve interstate commerce. Display fireworks, for example, which are purchased in one state and transported to another, require possession of a license from the Bureau of Alcohol, Tobacco and Firearms. Personnel of this bureau are empowered to investigate certain cases of arson, explosions, use of fire bombs, and careless activities relating to explosives. This function often brings these agents in contact with local and state fire department and law enforcement personnel. The Bureau also operates a highly sophisticated laboratory designed to aid in suppression of arson.

◆ U.S. DEPARTMENT OF THE INTERIOR

The Department of the Interior maintains two bureaus whose functions directly and indirectly contribute to public safety and fire prevention education.

NATIONAL PARK SERVICE

The National Park Service maintains a program to encourage conservation and environmental awareness in the national parks. This program, offered to the public, includes fire prevention as an environmental consideration. The exposure of the public to fire prevention measures and programs in the national parks undoubtedly has a residual effect outside of the parks.

BUREAU OF LAND MANAGEMENT

A good portion of the nation's federally owned lands in the western states are administered by the Bureau of Land Management. This agency provides fire prevention and fire suppression support in connection with lands it manages.

◆ U.S. DEPARTMENT OF AGRICULTURE

The federal department that has traditionally had the greatest fire prevention impact on the general public is probably the Department of Agriculture. The publicity campaigns of several of its agencies—the Forest Service, Natural Resources Conservation Service, and Extension Service—reach directly to most citizens.

FOREST SERVICE

The U.S. Forest Service of the Department of Agriculture maintains a widespread fire suppression force, and conducts an extensive nationwide fire prevention promotional program. This program, which has made Smokey Bear a national figure, reaches a high percentage of the U.S. population through television, radio, and newspaper advertising, as well as through signs and billboards.

Smokey observed his 50th birthday in 1994. His message, "Only YOU can prevent forest fires," is recognized by 95 percent of adults and 85 percent of children aged 8 to 12. The number of accidental wildfires caused by humans has been cut in half since Smokey arrived, despite the fact that the number of visitors to forests has grown tenfold during this period. The original Smokey was a cub that had survived a New Mexico wildfire.[2]

The Forest Service fire prevention program has been widely accepted by the public, as demonstrated by a number of polls. State forest fire personnel have cooperated in the federal efforts and have on many occasions promoted forest fire prevention before school groups and civic organizations.

CONSERVATION AND EXTENSION SERVICES

Closely related to fire protection is the program of the Natural Resources Conservation Service of the Department of Agriculture. This agency has been responsible for establishing countless farm ponds, which are put to use for fire protection by rural fire departments. Extension service programs include an emphasis on fire prevention. The National 4-H program, for example, includes farm fire prevention themes.

◆ U.S. DEPARTMENT OF COMMERCE

BUREAU OF THE CENSUS

A major function of the Bureau of the Census, an arm of the Department of Commerce, is to take a census of the U.S. population every ten years. Data collected may serve a variety of needs in the fire prevention field. For example, the census records may be used to identify the varieties of fire experiences within different types of communities and income levels. They may also be used for comparisons of effectiveness based on population densities.

NATIONAL INSTITUTE OF STANDARDS AND TECHNOLOGY

The National Institute of Standards and Technology has increased its role in fire prevention as a result of the enactment by Congress of the Fire Research and Safety Act of 1968.

The Institute, formerly known as the National Bureau of Standards, has a long history of cooperation with state and local governments in establishing standards of measurement and performance. As an example, the program of weights and measures, which greatly affects the marketing of all items sold on the basis of weight or other measurement, has long been under its purview. Although specific weight and measurement control programs are enforced at the state or local level, the overall standardization among states is based on procedures established by the National Institute of Standards and Technology.

The Institute's Building and Fire Research Laboratory is responsible for carrying out a great deal of research in the fire science, fire engineering, and fire safety fields. The Federal Fire Prevention and Control Act of 1974 strengthened programs previously operated by the Bureau in these fields.

Programs of the Building and Fire Research Laboratory are carried out by three divisions: the Materials and Construction Research Division, Building Environment Division, and Fire Research Division. The goal of the Fire Research Division is to enable engineered fire safety for people, products, facilities, and enhanced fire fighter effectiveness. The strategy to meet this goal is to reduce the risk of flashover in buildings, to advance fire safety technologies, to make advanced measurement, and to develop predictive models.

In addition to conducting fire-related research, the National Construction Safety Team (NCST) Act of 2002 authorized the Institute to establish an NCST team for deployment after events causing the failure of a building or buildings that resulted in substantial loss of life or posed significant potential for substantial loss of life. The Institute will lead a team of public and private sector fire and safety experts to conduct fact-finding investigations of building-related failures.[3]

Under the Occupational Safety and Health Act of 1970, the Department of Labor embarked on an enforcement program that brings it in close contact with a high percentage of places of employment within the United States. This act probably has more direct impact on fire prevention enforcement at the local level than any federal legislation previously enacted.

OCCUPATIONAL SAFETY AND HEALTH ADMINISTRATION

The regulations of the Department of Labor's Occupational Safety and Health Administration (OSHA) are applicable to employees, businesses, industries, institutions, farms, and other places of employment that engage in interstate commerce. The term *interstate commerce* includes all places that, for example, obtain supplies from, or market to, businesses in other states. For practical purposes there are few places of employment not covered by this broad definition. OSHA is charged with developing occupational safety and health standards and regulations. OSHA has promulgated a comprehensive set of regulations, including a number that incorporate National Fire Protection Association standards. Flammable liquids, liquefied petroleum gases, building exits, and electrical safety are among the fire protection subjects covered.

The legislation that established OSHA also established a procedure by which state governments can, if adequately prepared, enforce the provisions of the act within their jurisdiction. A number of states are utilizing this arrangement. In those that are not, the Department of Labor provides direct enforcement with its own personnel. Federal matching funds assist states that provide their own enforcement personnel. Local fire prevention inspections, as well as those of state agencies, are not preempted by this federal program, according to procedures set forth by the agency.

MINE SAFETY AND HEALTH ADMINISTRATION

The Mine Safety and Health Administration has a direct responsibility for fire prevention and safety within coal, metallic, and nonmetallic mines, often sharing this responsibility with a state bureau of mines. Records show that fires and explosions in coal mines have resulted in some major tragic losses of life. Inspection and fire prevention procedures in mines follow basic principles of fire prevention, but special consideration must be given to the unusual structural and atmospheric conditions inherent in mining operations.

◆ **U.S. DEPARTMENT OF HEALTH AND HUMAN SERVICES**

The Department of Health and Human Services has had a major impact on fire safety. This department, which was formed in 1953, has a number of agencies with direct interests in fire prevention.

HEALTH CARE FINANCING ADMINISTRATION

Responsibilities for ensuring the provision of proper fire safeguards in health care facilities receiving funds through Medicare (Title 18) and Medicaid (Title 19) rest with the Health Care Financing Administration. Fire prevention standards have been established for facilities housing patients in either program. Thousands of patients in long-term care facilities and hospitals are funded under these programs.

State agencies assist the Department of Health and Human Services in ensuring that fire safety requirements are met in these facilities. Funding may be discontinued for lack of compliance with the requirements, which encompass the Life Safety Code of the National Fire Protection Association.

◆ U.S. DEPARTMENT OF HOUSING AND URBAN DEVELOPMENT

One of the responsibilities of the Department of Housing and Urban Development is the development and direction of programs to bring about lower cost housing. The department has devoted considerable attention to fire protection for lower income housing units and has funded several research projects in the fire protection field. Requirements for manufactured housing are also enforced by this department.

FEDERAL HOUSING ADMINISTRATION

The Federal Housing program, an arm of the Department of Housing and Urban Development, has set certain standards for structures insured under the National Housing Act and other federal laws. These include minimum fire safety standards. Funding is provided by the agency for the construction of housing for older adults, nursing homes, intermediate care facilities, and nonprofit hospitals. Minimum fire protection standards are included in these mortgage insurance programs.

◆ U.S. DEPARTMENT OF TRANSPORTATION

The Department of Transportation, formed in 1966, includes a number of agencies that have an interest in fire prevention. Several of these agencies operated independently before the Department of Transportation was established.

FEDERAL AVIATION ADMINISTRATION

The Federal Aviation Administration is responsible for the control of all aspects of aviation, including safety features incorporated within aircraft. The agency ensures, for example, that carpeting in the cabin of passenger aircraft does not have an excessive flame spread. The agency is also concerned with the provision of adequate fire rescue forces and equipment at commercial aviation facilities.

FEDERAL HIGHWAY ADMINISTRATION

The Federal Highway Administration, through its motor carrier and highway safety sections, controls the safe transportation of dangerous cargoes, including explosives and flammables, in interstate commerce over highways. This agency conducts inspections to ensure compliance with regulations covering safeguards for such movements. All commercial carriers employing motor vehicles come under this program when their operations extend to interstate commerce. The motor carrier and highway safety sections are responsible for the investigation of accidents involving controlled carriers.

FEDERAL RAILROAD ADMINISTRATION

The Federal Railroad Administration is responsible for the regulation of railroads. The Bureau of Railroad Safety has specific responsibility for enforcing regulations on safety and fire prevention on interstate railroads.

RESEARCH AND SPECIAL PROGRAMS ADMINISTRATION

The Research and Special Programs Administration (RSPA) coordinates Department of Transportation responsibilities pertaining to pipeline safety and the transportation of hazardous materials. Responsibilities include the administration of safety regulations, enforcement, and issuance of exemptions and interpretations.

RSPA's Office of Pipeline Safety establishes and provides for enforcement of safety standards for the transportation of hazardous and gaseous materials by pipeline. In the pipeline safety program, states may be awarded up to 50 percent of their costs in carrying out enforcement and inspection programs.

RSPA is the Department of Transportation's focal point for regulations, exemptions, research, coordination with the states, and the emergency response information published in its guidebook. Cooperation with state agencies is especially important if the state itself incorporates Department of Transportation regulations within its motor vehicles code, which a number of states do.

The North American *Emergency Response Guidebook* is the most widely distributed technical guide for initial response actions in the event of incidents (spill, explosion, fire) involving hazardous materials. This publication is distributed to emergency responders through a network of key state agencies. It is cosponsored by the Canadian and Mexican governments.

◆ U.S. DEPARTMENT OF VETERANS AFFAIRS

The Department of Veterans Affairs, although not directly reaching the public at large with respect to fire prevention, has performed work that is of value in the field. Extensive research by the Veterans Administration in the use of noncombustible bedclothing is looked on as a progressive step, and fire prevention criteria developed by the Department are used as guidelines by many nongovernmental hospitals.

◆ U.S. DEPARTMENT OF HOMELAND SECURITY

The Department of Homeland Security became a cabinet agency in 2002. It was established as a direct result of the events of September 11, 2001, bringing under one secretary a number of responsibilities of critical interest to the fire service. A major funding initiative is available through this agency on a matching grant basis. These support functions relate to emergency response and firefighter safety as well as fire prevention and emergency communications. In addition several agencies with direct responsibilities for fire service support and liaison are housed in the Department.

Legislation enacted to create the Department of Homeland Security and subsequent presidential directives have enhanced the role of the federal government in preparation for and response to acts of terrorism and other emergencies.

COAST GUARD

The U.S. Coast Guard, a part of the new department, has long had a legal responsibility in the field of fire prevention. Under the port security program of the Coast Guard, regulations pertaining to fire prevention and other security features are enforced in the ports of the United States. The Coast Guard has responsibility for inspection and regulatory enforcement on all waterfront facilities. This function means that a Coast Guard inspector has concurrent jurisdiction with municipal and state fire prevention agencies. The regulations enforced include requirements for fire extinguishing equipment, marking of extinguishers, maintenance of access for fire fighting equipment, control of cutting and welding operations, and control of smoking.

The Coast Guard also has a fire safety function in connection with vessels operating in U.S. waters. Vessels constructed in American shipyards are under the constant inspection of the Coast Guard during construction. A major portion of the obligatory inspection relates to fire safety. Once the vessel is commissioned, the Coast Guard continues to inspect and ensure proper maintenance of fire fighting equipment and training of crews.

The loading and storage of cargo, especially that of a hazardous nature, is also under the purview of the Coast Guard. Coast Guard personnel may be detailed to directly monitor the loading and stowage aboard ship of explosives and other dangerous commodities. Comprehensive regulations relating to the transportation and the storage of dangerous articles of all descriptions are enforced by the Coast Guard.

Under the provisions of the Federal Boating Act of 1958, Coast Guard boating safety teams are required to inspect small boats to ensure compliance with required safety measures, including fire safety. The Coast Guard also has responsibilities in the field of abatement of oil pollution and other environmental issues.

In addition to surveillance over vessels, the Coast Guard is responsible for ensuring that merchant marine personnel have qualified through training. Licenses are issued to such personnel for the performance of specific duties after satisfactory completion of a comprehensive examination. These examinations, both written and practical, include a number of fire safety problems.

The Coast Guard also conducts investigations into causes of marine disasters, including fires. Appropriate sanctions are imposed where indicated.

BUREAU OF CUSTOMS AND BORDER PROTECTION

This Bureau, which is also part of the Department of Homeland Security, maintains control over merchandise, including fireworks, imported from other countries. It maintains representatives in all ports of entry. Bureau agents determine that markings for flammability, for example, appear on imported textile products as required by the federal labeling provisions. They are also active in the prevention of smuggling.

FEDERAL EMERGENCY MANAGEMENT AGENCY

This agency, a part of the Department of Homeland Security, is the federal government's focal point for emergencies. In addition to the U.S. Fire Administration, which includes the National Fire Academy, the agency is responsible for emergency management and other coordinating programs relating to major disasters.

The Fire Research and Safety Act of 1968 provided for fire research activities and for the establishment of the National Commission on Fire Prevention and Control. The commission had the task of recommending to the president and to Congress procedures for reducing fire losses in the United States.

After extensive study, the commission submitted a report entitled *America Burning* to the president and Congress. This report, issued in 1973, contained as a major recommendation the establishment of "an entity in the Federal Government where the Nation's fire problem is viewed in its entirety, and which encourages attention to aspects of the problem which have been neglected."

Needs that the Commission felt could be addressed by such a federal agency included more emphasis on fire prevention, better training and education in the fire service, more education of the public about fire safety, and recognition of the hazardous environment to which Americans are exposed from a fire safety standpoint. They also addressed the need for improvements in building fire protection features and in research activities related to fire safety.

Congress subsequently enacted legislation in 1974 establishing the National Fire Prevention and Control Administration as an agency of the Department of Commerce. Under the provisions of President Carter's 1979 reorganization plan, the National Fire Prevention and Control Administration was transferred basically intact to the new Federal Emergency Management Agency. The National Fire Prevention and Control Administration became the U.S. Fire Administration under separate legislation.

The National Fire Academy, which is housed with the Emergency Management Institute at the National Emergency Training Center in Emmitsburg, Maryland, is responsible for advanced-level fire and emergency services education activities. A wide variety of programs is offered on campus and in field classes offered throughout the country. This academy has gained international stature and is recognized as having a major impact on fire service activities in the United States. Both volunteer and career fire service personnel can take advantage of course offerings. A number of programs relate to public fire education, code enforcement, arson suppression, and fire prevention administration.

The agency has responsibilities in the implementation of the Hotel and Motel Fire Safety Act of 1990 under which federal employees are to stay in properties meeting the law's fire protection requirements.

The Fire Administration also operates the National Fire Incident Reporting System. This system, which is described in more detail in a subsequent chapter, has enabled fire safety personnel to study the causes of fires and to develop more effective public education and legislative programs.

Juvenile fire setters have been the focus of another program of the Fire Administration. Fire service personnel have been trained in methods aimed at reducing such fires through effective counseling and education.

From time to time, the Fire Administration has conducted special studies of major fires and has looked at specific fire problems in high-risk areas. Efforts have been made to disseminate information on these matters to bring about improved community fire safety.

A program called *Partnerships in Fire Safe Building Design* is being offered by the U.S. Fire Administration in an effort to bring building design professionals and code enforcers together to build understanding of their common ground. Fire Administration officials see this as a means of reducing variations in interpretation of similar codes in different jurisdictions.[4]

◆ INDEPENDENT U.S. GOVERNMENT AGENCIES

The Consumer Product Safety Commission has endeavored to eliminate hazards associated with consumer products, including toys, hazardous substances, and flammable fabrics. The agency has an enforcement and educational program that strives to reduce injuries resulting from use of common consumer products, most of which are transported through interstate commerce. The agency enforces regulations requiring proper labeling of hazardous substances and is authorized to legally ban from the market commodities that pose severe fire or safety hazards. It enforces the Flammable Fabrics Act as well. This law prohibits interstate marketing of wearing apparel and other materials that do not conform to flammability standards.

The Consumer Product Safety Commission has been active in the field of fireworks control and has banned the manufacturing and distribution of the more explosive fireworks. Flammable liquids for household use have also received the scrutiny of the Commission. Enforcement officers for the agency seek out items having unusual characteristics from a safety standpoint. The activities of this agency are having an increasing impact on fire safety.

The Environmental Protection Agency has the responsibility of ensuring the protection of the environment through the abatement and control of pollution. The comprehensive charge to this agency, organized in 1970, includes development of standards for open-air burning and a number of other fire-related pollution problems.

The Federal Communications Commission, although not as directly involved in fire prevention as some other agencies, has served the cause of fire prevention by its mandatory requirements for public service time on radio and television broadcasting stations. Many fire departments have taken advantage of this provision and have prepared public education fire prevention programs for television and radio. The agency also has contact with the fire service through control of radio frequencies used by public safety agencies.

The Federal Trade Commission administers regulations pertaining to advertising that have been helpful in eliminating false descriptions of products that may be dangerous or that do not perform properly. In several cases, the commission has acted against the manufacturers of small throwaway fire extinguishers to which labels indicating superior performance had been attached.

The General Services Administration serves as a management agency for the construction and operation of government-used buildings, the procurement of supplies, and the control and disposal of records. In carrying out these responsibilities, fire protection standards have been established by the Public Buildings Service, an arm of the General Services Administration.

The National Aeronautics and Space Administration has made valuable contributions to fire prevention and public safety in connection with space research. Highly reliable fire-retardant materials developed for use in spacecraft have been adapted for commercial purposes and are available in a number of products on the market.

The National Institute of Building Sciences was established by Congress in 1978 for the purpose of providing "an authoritative nationally recognized institution—to advise both the public and private sectors of the economy with respect to the use of building science and technology.[5] As a quasi-governmental agency, the Institute depends on private as well as federal support. Board members are selected by both groups, the president appoints 6 of the 21 members, and the remaining positions are selected by the private sector.

The Institute has been quite successful in getting groups with diverse interests to work together in the common causes of improved building safety and more economic construction costs. It also advises other federal agencies on matters relating to building construction. Fire safety has been a high priority of the Institute.

The Nuclear Regulatory Commission regulates and licenses the civilian use of nuclear material and generally has a responsibility in connection with the safety of this material. Fire prevention procedures are a factor in these regulations.

The American Red Cross, another quasi-official agency, has joined in national fire prevention campaigns and actively supports programs for the reduction of fire losses.

The National Transportation Safety Board has specific statutory responsibility for the investigation of accidents in civil aviation as well as highway, rail, marine, and pipeline accidents. Investigation is mandatory in aviation accidents; other accident investigations are made on the initiative of the board.

Members of the board, who are appointed by the president, have executive authority in conducting their investigations and also have the authority to conduct formal proceedings for review of appeal on the suspension, amendment, modification, revoking, or denial of any certificate or license issued by the secretary of transportation or by an administrator operating with the department. The Safety Board's reports of its investigations of major accidents have resulted in changes in regulations and in changes in procedures within the Department of Transportation.

◆ STATE AGENCIES

All state governments have agencies with responsibilities in the field of fire prevention. Code enforcement and inspections at the state level have been responsible for a multitude of improvements in fire safety measures and fire prevention practices.

STATE FIRE MARSHAL

The state agency normally having fire prevention as a major responsibility is the state fire marshal's office. Forty-nine states now have an organization carrying out fire safety functions at the state level; such an agency does not exist in Hawaii. All provinces and territories in Canada have such offices.

State fire marshals' offices were first established shortly before 1900. Since that time they have grown in number at a rather steady rate. Some states have established fire marshals' offices as a direct result of some major tragedy; others have done so because of a realization that effective fire prevention must be carried out as a function of the state government.

In several states, the state fire marshal's office is an arm of the state police. The agencies carry out a full range of fire marshal's duties, including fire prevention, code

FIGURE 10.1 ◆ This marker in Texas City gives sobering reflection on the tremendous death toll in that 1947 tragedy.

enforcement, and fire investigation; however, responsibilities in the inspection and code enforcement field may be limited, with other agencies assigned the remaining inspection responsibilities in a portion of these states.

Because of the past close association of fire prevention with the insurance industry, the office of the fire marshal in some states is assigned to the office of the insurance commissioner. Another reason for this close relationship with the insurance department is that in some states the combined office of the insurance commissioner and fire marshal is funded by a tax on insurance premiums.

In an increasing number of states, the office of the fire marshal has become a part of the state department of public safety, an arrangement that makes possible closer cooperation with other state law enforcement agencies, especially in the field of fire investigation.

Personnel Appointments. Considerable diversity exists in the ways fire marshals are selected. In some states, the position is filled by an elected official or by gubernatorial appointment; in others, appointment is made by the state fire board or fire prevention commission, or by the attorney general, state comptroller, or other person under whom the fire marshal's office operates. Length of tenure varies from an indefinite period to four, five, or six years. In states in which the fire marshal's office is an arm of the state police, the appointment is usually based on departmental assignment procedures. Alabama's fire marshal is a merit system position.

Appointment methods for deputies and other personnel within the fire marshal's office are also varied. In some states, all deputies are merit system or civil service employees; in others, the deputies are appointed to serve at the pleasure of the fire marshal. In the latter situation, suggestions on appointments may come from someone at a higher level in the state government, with possibilities of political involvement in the appointments. This was fairly common practice in past years. Now, however, in a majority of the states, public employees are selected under a merit system or other nonpatronage employment program.

Fire Marshal's Responsibilities. Although the responsibilities of the office of the fire marshal vary from state to state, its primary function is usually to enforce state fire prevention codes and to investigate suspicious fires. The office is also normally the coordinating agency for all fire prevention activities within the state government. The state fire marshal strives to obtain compliance with fire prevention regulations in connection with the licensing program of other state agencies, such as the health department and department of education. The fire marshal may recommend the inclusion of fire safety requirements in licensing and accreditation procedures and in other state controls imposed on facilities within the state.

Fire Code Enforcement. In a number of states, the office of the state fire marshal is the primary fire code enforcement agency. In some states, a locally designated fire marshal or fire chief may serve as an ex officio deputy to the state fire marshal for the purpose of code enforcement within the local jurisdiction.

Fire code enforcement at the state level may on occasion have advantages over strictly local enforcement. Local political pressures may be less effective in hampering fire prevention code enforcement administered at the state level than at the local level. The state fire marshal's office can serve as a backup for the local fire marshals' offices, to render assistance and support in difficult situations. The state fire marshal's office can also lend support in the form of technical advice and services to smaller communities, whose budgets do not usually allow for specialized personnel.

Fire Investigation. The state fire marshal's office has the responsibility in most states to conduct investigations of all fires to which it is called. Notification procedures vary; however, local fire and police departments are usually the source of calls.

Under the Model Fire Marshal Law,[6] which is enacted in many states, the state fire marshal or a representative of the office has right of entry for the investigation of fires and responsibility for control and suppression of arson within the state. A number of states invest the fire marshal with powers of subpoena and arrest; under these legal conditions, the office is considered to be a law enforcement agency within the framework of state government. The state fire marshal's office may provide investigators for investigation of suspicious fires. The agents are usually qualified as expert witnesses in the courts and can play a major part in the successful prosecution of a case. In many states, fire marshal's office personnel are considered peace officers and are provided with firearms. These individuals have the power of arrest in arson cases.

Fire Statistics. Another duty carried out by most fire marshals' offices is the compilation of fire statistics. An effort is being made to encourage every state to engage in fire recording and to develop statewide statistical data for use in fire safety programs and legislative hearings.

Several states have statutes requiring fire departments to record all fires within the state; in others, such an obligation is also placed on insurance carriers doing business in the state. Certainly, it is necessary to have reports from both fire departments and insurance carriers to develop an accurate statistical compilation of fire losses and insurance data. A review of reports submitted by insurance companies will reveal many losses that were not reported to the fire department.

Fire and Life Safety Education. A number of state fire marshals operate public fire and life safety education programs. In some cases, state efforts are directed toward the development of local capabilities in this field; in others, fire marshal's office personnel provide programs directly to the public. Ohio's fire marshal has long been a leader in this field. His program is described here:

Fire Prevention Bureau
Ohio Division of State Fire Marshal
The Ohio Division of State Fire Marshal, created in 1900, is among the longest serving state-level fire marshal's offices. The initial duty of the Fire Marshal was to investigate and record all fires in the State. Within a couple of years, the Fire Marshal was also charged with creating a Fire Prevention Bureau to address the State's growing "fire problem." Throughout the next century, the Fire Marshal's fire prevention efforts greatly expanded. Ten Fire Safety Educators annually provide about 3,500 fire and life safety training and education programs throughout the State for nearly 80,000 individuals. Major program emphasis is on school children, older adults, and health care facilities.

The Bureau has also developed two dozen topical fire safety pamphlets and other materials, including such items as home fire safety planning (EDITH), smoke and carbon monoxide detectors, and home and business safety checklists. Over a half million items are distributed each year to Ohioans. Project SAFE (Smoke Alarms For Everyone) provides smoke detectors to fire departments in high-loss areas of the State to install in residences. A Smoke DOG (Detectors On Guard) Award program recognizes citizens who take the proper actions when fire strikes their home. The Bureau also supports the fire safety training and educational efforts of local fire departments, maintains a Juvenile Fire Setter intervention program, and maintains a website with program materials, information and fire statistics.[7]

Other Duties. A number of state fire marshal's offices have been given the responsibility for enforcement of manufactured housing regulations. Georgia and Tennessee have comprehensive programs in this field. Although these controls relate to fire safety features, plumbing, sanitation, structural stability, and other phases of construction are also included. This means that the fire marshal's office must have an ability to handle these related areas as well. The responsibility for inspecting all electrical work is assigned to the fire marshal in some states, including Tennessee and Wyoming.

Explosives control activities are assigned to the fire marshal in some states. This control may include licensing programs for people possessing and using explosives, as well as a capability to deactivate clandestine devices, bombs, and other explosive devices.

The advent of greater federal participation in fire safety has brought about an interest in improved coordination of these services at the state level. Among concepts advanced is that of a state coordinating commission or council whose purpose is to influence fire safety efforts in the state. The council or commission may or may not have direct authority over any operating agencies, such as the fire marshal's office or state fire training program. Such a commission would likely have as representatives fire chiefs, both career and volunteer firefighters, building officials' associations, the state fire marshal and director of fire service training, the state forester, and members of other statewide agencies and organizations that have responsibilities and interests in fire protection.

FIGURE 10.2 ◆ The size of this door portrays the importance some fire services place on fire prevention.

HEALTH DEPARTMENT

All 50 states have a department of health in the state government. The health agency has an interest in fire safety as a part of its responsibility in safeguarding citizens from health hazards and, more specifically, in ensuring proper safeguards for facilities under the direct control of the agency: hospitals, extended-care facilities, and other such institutions. The health department may also have an interest in fire prevention through accident prevention programs, which in most states include certain fire safety information.

In view of the immobile nature of many occupants, institutions under health department control have been required to upgrade fire protection to ensure adequate safety for patients. The state health department is in a position to exert strong influence in the field of fire safety. The work of the sanitarian, for example, in checking the cleanliness of a restaurant has a bearing on fire safety. A clean restaurant is less likely to invite fire damage.

EDUCATION DEPARTMENT

Each of the 50 states has a department of education. This agency has an interest in ensuring the safety of all schools within the state, including those that are privately operated but come under state accreditation control. As a rule, state department of education regulations require periodic fire inspection and plan reviews for new construction, evacuation plans, fire drills, and other procedures to strengthen fire protection and fire prevention. In some states, school fire safety regulations are issued and enforced in cooperation with the office of the state fire marshal.

Departments of education may also be responsible for including fire prevention education in the required curriculum. These departments encourage fire services personnel to prepare children for lifelong awareness of the importance of fire prevention.

DEPARTMENT OF LABOR AND INDUSTRY

State governments have a division for regulation of labor and industry. Although primarily responsible for industrial safety, these agencies often include fire prevention in their enforcement and educational programs. With the advent of the Occupational Safety and Health Act of 1970, the responsibilities of the department of labor and industry have increased in many states. In a number of states, this agency has been designated by the governor as the primary enforcement and coordinating agency for state enforcement of the provisions of OSHA.

In several states—Pennsylvania and Wisconsin, for example—in which the fire marshal's duties do not include a full range of fire code enforcement, inspections relating to fire prevention are the responsibility of the department that regulates industrial safety. Likewise, the control of explosives may be in this agency.

Industrial safety classes organized and sponsored by departments of labor and industry may include emphasis on fire prevention. North Carolina has a comprehensive industrial safety education program that stresses fire prevention as an element of safety.

A number of states have enacted "right to know" laws. Under these laws employers are required to furnish their employees information on toxic substances in the workplace and are required to provide training on the safe handling of such substances in emergency procedures. Employers must notify local fire departments of the location and characteristics of all toxic substances regularly present in the workplace. This legislation may affect fire prevention bureau activities since much of the information relates to that sought during code enforcement visits.

Federal legislation likewise requires that information relating to toxic materials be given to employees and to the local fire services under the Superfund Act of 1987. Fire safety efforts are enhanced by greater distribution of information relating to hazardous materials.

PUBLIC UTILITIES COMMISSION

The public service or public utilities commission is another state agency that has an interest in fire prevention. These commissions normally regulate public utilities, including gas and electrical distribution. Both of these utilities have a bearing on fire safety. The public service commission in some states regulates transportation of hazardous commodities. Again, a function having some bearing on fire prevention is exercised by the commission.

In several states the public service commission has a statutory responsibility for conducting electrical inspections within the state or for seeing that such inspections are conducted properly. Electrical inspections are most essential in any comprehensive fire prevention program.

RACING COMMISSION

Although not found in every state, the racing commission is another agency that may have some influence in the field of fire prevention. Requirements for fire safety may be imposed by the commission as a condition of continued state recognition of racetracks. An interesting feature of code enforcement is raised in this phase of activity. To ensure continued operation of a track, steps must be taken to provide a greater degree of life safety for horses than would be required for animals under the fire safety code in effect in the state. The economic impact of the racing industry is a factor.

BUREAU OF MINES

States having operating mines generally have a state agency responsible for control of the industry. The bureau of mines usually includes fire prevention as a part of its legally constituted duties. Fire prevention and protection in underground mines is a field quite remote from structural fire prevention.

FORESTRY DIVISION

Forest fire prevention is a major function of divisions of forestry in all states and provinces. These agencies are generally responsible for fire prevention and control in their areas of jurisdiction. They are normally established in such a way as to permit assignment of personnel to fire prevention or suppression, depending on immediate demands. During fire seasons, for example, all personnel are assigned to fire suppression duties, whereas when fires are less frequent, these employees may be used for other duties.

Fire prevention efforts of state forestry departments have undoubtedly been responsible for the dissemination of fire safety information to countless individuals who would not otherwise have been contacted.

OTHER STATE AGENCIES WITH RELATED FUNCTIONS

A number of state agencies have functions that are less directly related to fire prevention. A rather obscure example is the agency that examines and licenses barbers.

By including fire prevention as a requirement for licensure, the agency can bring about improved fire prevention practices in this type of occupancy. Although not a high life-hazard occupancy, the effectiveness of state backing for fire code enforcement can be helpful to local fire prevention officers.

The functions of several other state agencies in this category are described in the following paragraphs:

State Medical Examiner. Although perhaps not thought of in this regard, the state medical examiner or postmortem examiner actually performs a function having a bearing on fire prevention. The persons assigned to such an agency are responsible for performing autopsies to determine the cause of death and other related factors.

All fire deaths should be investigated, and an autopsy should be performed. Through this means, a factual picture of our fire prevention problems as related to life safety can be more readily obtained. In one state, for example, where such a program was vigorously implemented, it was learned that a number of deaths from causes other than fire were being reported as fire deaths. Individuals in some cases had been killed by weapons before the setting of the fire. Had autopsies not been performed, the deaths would have been carried as fire fatalities. However, a body damaged by fire is not pleasant to examine, and in states where nonprofessional coroners are used, there may be a tendency to automatically write the death off as "smoking in bed" or some other fire-related cause when in fact the victim died from a cause unrelated to the fire. Another responsibility is that of determining whether the victim died from carbon monoxide poisoning before being burned in the fire.

Commissions on Aging. Most states have an agency responsible for problems of concern to older adults. These agencies can be of great assistance in ensuring the dissemination of fire prevention information to older adult citizens. Thousands of older adults who are not living in institutions are contacted on a regular basis by these agencies.

State Institutions. State agencies with responsibilities for housing patients, inmates, and other wards of the state are likewise responsible for fire prevention measures in connection with the operation of their facilities. Through the cooperation of inspecting agencies, such as the fire marshal's office, fire safety practices in these institutions can be maintained at a high level.

Mental health agencies often have responsibility for licensing structures housing people suffering from various mental disorders. The provision of fire safety requirements will have a bearing on the safe operation of the facilities.

Gaming Commission. A number of states have gaming commissions, which regulate casinos, riverboats, and other gambling operations. Usually, these agencies require strict adherence to fire safety requirements in gaming facilities. As an example, when South Dakota made gambling legal in Deadwood, the state fire marshal was given the responsibility for requiring many fire safety improvements in existing buildings.

Building Code Agencies. Many states have enacted a state building code on either a mandatory or permissive basis, with functional responsibility in a variety of state agencies. Under the provisions of some state laws, a local jurisdiction may, if it deems it desirable, adopt the state building code for enforcement within the jurisdiction.

Educational Institutions. The college and university system operated by state governments has a very definite role in fire prevention in two kinds of programs: those conducted for the safety and well-being of students residing on campuses and occupying classroom spaces; and fire service training programs to improve the public fire departments.

An example of state university fire safety outreach is the smoke detector initiative of the University of Alabama at Birmingham Injury Control Research Center. This project seeks to have smoke alarms in every Alabama residence. The state fire marshal, state department of health, fire departments, and community action groups are cooperating in this program.

State fire training programs operated by colleges and universities are among the most effective in the fire service training field. By being housed on campuses, they have the advantage of the tie-in with academic programs and of offering participation in research work that might not otherwise be available. Foremost among these fire service training programs are those conducted by Louisiana State University, the University of Maryland, Oklahoma State University, and Texas A&M University. A number of colleges and universities offer four-year and advanced degree programs in fire protection. In some states, fire service training programs are operated as part of the state department of vocational education.

Judicial Branch. The judicial branch of state government has a bearing on fire prevention through legal processes in fire prevention code enforcement and arson control. Legal action in code enforcement is effective only with the backing and support of the judicial branch of government.

A recent Ohio Supreme Court case is an example. The court held that the use of adapters to connect hose to fire hydrants because of thread differences was not a violation of Ohio statute since there was no support for the argument that the use of adapters created potential problems for the fire department. A local fire chief had alleged that since fire hydrant threads used by a county water system were dissimilar to those in his jurisdiction, necessitating the use of adapters, a noncompatibility situation was created.[8]

Legislative Branch. The legislative branch of state government must be mentioned because of the major role it plays in enacting legislation relating to fire prevention, fire and arson control, and its role in enacting budgets for fire prevention and fire protection.

◆ COUNTY AGENCIES

Fire safety activities at the county level are quite varied from one county to another.

An increasing number of county governments are operating fire departments for the protection of the entire county, or at least for portions outside of incorporated cities. The Los Angeles County Fire Department protects unincorporated areas as well as a number of incorporated cities with which it contracts to provide fire protection. Fire prevention services are also provided.

Although some metropolitan counties provide a full range of services and activities similar to those normally associated with municipal governments, the majority of the counties in the United States are not heavily involved in fire prevention programs.

Many counties provide fiscal support for volunteer fire department operations and some maintain fire marshals' offices, communications centers, and training programs to assist local fire departments.

Because so-called full service counties are rather rare, the programs they offer are considered in the following discussion of municipal government activities. An ever-expanding number of counties are enforcing planning and zoning requirements at the county level.

◆ MUNICIPAL GOVERNMENT

Municipal government has a tremendous impact on fire prevention and fire control. There are few municipalities, regardless of area or population, that do not have some provisions for public fire protection in the form of an organized fire department, which may have career, volunteer, or both types of personnel. Details of municipal government's role in fire prevention are delineated in Chapter 7.

◆ SUPPORT AGENCIES

Certain agencies operating at all levels of government have what may be called an indirect association with fire prevention. Government organizations responsible for taxation and assessment and for fiscal operation of government at all levels provide important support for effective fire prevention. Their part in ensuring funding of necessary programs cannot be taken lightly.

Agencies such as emergency management, art commissions, transportation, law, beautification groups, local and state environmental agencies, historical preservation boards, and a number of others have some limited involvement with fire prevention and should not be overlooked for community cooperation.

◆ SUMMARY

Government, except at the municipal level, has not been as active in fire prevention and protection in the United States and Canada as it has been in other countries. However, with population growth and economic changes in the 20th century bringing about concentration of the population in urban areas, the influence of governmental agencies in fire prevention efforts at all levels is steadily increasing. Indicative of this trend is the creation by Congress in 1973 of the National Fire Prevention and Control Administration, now known as the U.S. Fire Administration, a part of the Department of Homeland Security.

Among federal agencies, the Department of Agriculture has traditionally had the greatest impact on the general public with respect to fire prevention (mainly through the Forest Service). However, the National Institute of Standards and Technology and the Federal Emergency Management Agency, an arm of the Department of Homeland Security, have made many contributions in the field.

Under the Occupational Safety and Health Act of 1970, the Department of Labor has an impact on fire prevention through the enforcement procedures of the act, which apply to businesses, industries, institutions, farms, and other places of employment that engage in interstate commerce.

The Department of Transportation is also an important force in fire prevention. The Department of Defense has maintained fire prevention efforts and training for members of the armed forces, which carry over into civilian life.

Among state agencies, the office of the state fire marshal is paramount, having wide responsibilities in fire code enforcement and inspections, and investigations of suspicious fires, among other duties. The state fire marshal also lends aid and support to municipal fire departments.

At the municipal level, the community fire department is without question the major bulwark against the perils and ravages of fire in the United States. Practically every community has such a force, whether it is career, volunteer, or a combination of the two.

■ ■

Review Questions

1. The fire safety of U.S. embassies is the responsibility of what department?
   a. Treasury
   b. State
   c. Defense
   d. Justice
   e. Commerce
2. The United States Army has contributed to the field of fire prevention through
   a. development of fire-retardant materials
   b. studies of hazardous materials
   c. Class A foam
   d. all of the above
3. The FBI can help fire prevention by
   a. arresting arsonists
   b. taking fingerprints
   c. examining materials found at the fire scene
   d. all of the above
4. Personnel of the Bureau of Alcohol, Tobacco and Firearms can investigate
   a. certain arson cases
   b. explosions
   c. use of fire bombs
   d. all of the above
   e. none of the above
5. The National Park Service programs include
   a. Smokey Bear
   b. fire prevention
   c. wildland interface
   d. all of the above
   e. none of the above

6. The number of accidental wildfires caused by humans has been reduced by _____ since Smokey Bear arrived.
   a. 10 percent
   b. 20 percent
   c. 30 percent
   d. 40 percent
   e. 50 percent
7. The Federal Highway Administration
   a. investigates all accidents
   b. investigates accidents on federal highways
   c. investigates accidents involving controlled carriers
   d. all of the above
   e. none of the above
8. The Department of Homeland Security was created in
   a. 1999
   b. 2001
   c. 2002
   d. 2003
   e. 2004
9. The National Fire Academy is in
   a. Washington, D.C.
   b. New York, New York
   c. Chicago, Illinois
   d. Denver, Colorado
   e. Emmitsburg, Maryland
10. State fire marshals exist in all states except
   a. Hawaii
   b. Alaska
   c. Arizona
   d. Mississippi
   e. North Dakota

Answers

1. b	6. e
2. b	7. c
3. d	8. c
4. d	9. e
5. b	10. a

Notes

1. U.S. National Commission on Fire Prevention and Control Report, *America Burning* (Washington D.C.: U.S. Government Printing Office, 1973).
2. "Smokey Bear Turns 50," *Parade* (February 6, 1994).
3. Personal communication to the author from Director, National Institute of Science and Technology Building and Fire Research Laboratory, Gaithersburg, Md., 2003.
4. U.S. Fire Administration, Memorandum (Emmitsburg, Md., 2003).
5. National Institute of Building Sciences, *Annual Report to the President of the United States* (Washington, D.C., 1997).
6. *Model Fire Marshal Law* (Quincy, Mass.: International Fire Marshals Association of America).
7. Personal communication to the author from Greg Drew, Chief, Fire Prevention Bureau, Ohio State Fire Marshal's Office, 2004.
8. *Gamble et al. v. Dobrasky* (Supreme Court of Ohio, Case 99-1311, 2000).

Fire Prevention through Arson Suppression

CHAPTER **11**

Suppression of arson is clearly a factor in fire prevention efforts. In many communities, arson is a major part of the fire problem; therefore, any steps that can be taken to prevent arson reflect favorably on the community's fire statistics.

Although law enforcement agencies often share with fire services the responsibility for suppressing arson, the latter service is usually the prime mover in arson control. Not only can arson be discouraged by effective investigation and prosecution, but many communities have also mounted successful programs aimed at preventing such incidents. In this chapter, various aspects of arson control are discussed; however, this is not intended as a complete treatise on the subject.

◆ THE CRIME OF ARSON

As a type of criminal activity, the maliciously set fire historically has been considered a serious crime. As an example, the first arson law in Maryland, enacted by the General Assembly meeting in St. Mary's City in 1638, classified arson as a felony. Punishment was death by hanging, loss of a hand, or being burned on the hand or forehead with a hot iron. The offender also forfeited ownership of all properties. If the offender escaped death for the first offense, there was a mandatory death sentence for the second offense.[1]

In Illinois, which obtained statehood in 1818, arson was considered so serious an offense that between 1819 and 1827 it was one of only four crimes punishable by death.[2]

Early western towns were likewise hostile toward arsonists. Vigilante justice occasionally superseded the established system. Closely built frame buildings, coupled with meager fire suppression capabilities, could spell doom for the entire town when an arsonist was at work.

A suspected arson ring in Virginia City, Nevada, was dealt with by a vigilante band that stormed the local jail in 1871 and removed and lynched a suspected arsonist. Other members of the arson ring were sentenced to the state prison. Arson did not occur frequently after that time.[3]

FIGURE 11.1 ◆ This Cottonwood Falls, Kansas, Court House is typical of older public buildings that are subject to arson fires. Many are being sprinklered during restoration.

Arson is one of the few criminal activities in which there is often no immediate recognition that a crime has been committed. A fire is considered to be an accident unless a thorough investigation shows evidence of arson. The civil disorder fire is a different matter. In this type of maliciously set fire, it is readily apparent in most cases that a crime has been committed. But the investigator must still prepare a tight case that examines as far as possible ways in which the fire could have been started accidentally.

The crime of arson has another unusual feature. The weapon most often used is readily available. This weapon is the ordinary match, which is, of course, available at most every convenience store, service station, and restaurant. No program could be developed to control the acquisition of matches, nor would such a cause have popular support.

ARSON STATISTICS

The *Uniform Crime Reports,* published by the Federal Bureau of Investigation of the U.S. Department of Justice, provides interesting statistics on arson in the United States. Statistics for 2002 are used in this chapter; however, a study of statistics for other recent years will indicate that there is a fairly constant pattern in arson statistics for the past five or six years.[4]

A total of 66,308 arson offenses were reported to the Uniform Crime Reporting (UCR) Program as occurring in 2002. Arson is defined by the Program as any willful or malicious burning or attempt to burn, with or without intent to defraud, a dwelling house, public building, motor vehicle or aircraft, personal property of another, etc.

Only fires determined through investigation to have been willfully or maliciously set are classified as arsons. Fires of undetermined origin are excluded.

The arson rates ranged from 52.3 per 100,000 inhabitants in cities with populations over 1 million to 16.6 per 100,000 rural county inhabitants. The suburban counties and all cities collectively recorded rates of 27.0 and 36.5 per 100,000 inhabitants, respectively. Overall, the 2002 national arson rate was 32.4 per 100,000 population.

A review of the statistics regarding incendiary and suspicious U.S. building fire losses published by the National Fire Protection Association shows a 61 percent decrease from 1978 to 2001. Incendiary and arson fires remain the number-one cause of property damage due to fire in the United States. In 2001, for the eighth straight year, juvenile fire setters accounted for a majority of those arrested for arson.

The UCR Program identifies arson by type of property burned. In 2002, structural arsons (residential, commercial, industrial, etc.) were the most frequent type, representing 41.3 percent of those arsons. Mobile properties (motor vehicles, trailers, etc.) were 33.1 percent of arsons, and other properties (crops, timber, etc.) accounted for 25.7 percent of arson offenses.

Residential properties accounted for 60.7 percent of those offenses. Of residential occupancies, single occupancies comprised 71.0 percent of the total residential arsons. Approximately 18.2 percent of structural properties were uninhabited or not in use. Motor vehicles comprised 94.6 percent of the total mobile properties.

In 2002, the average dollar loss of arson offenses was $11,253. Structural property arsons had an average dollar loss of $20,818; mobile property arson offenses were $6,073; and other property type arson offenses were $2,536.[5]

The National Fire Protection Association reported that intentionally set fires in structures resulted in 2,781 civilian deaths in 2001; 2,451 due to the events of September 11, 2001, and 330 in other set structure fires. Intentionally set structure fires also resulted in $34.453 billion in property loss: $33.44 billion due to the events of September 11, and $1.013 billion in other set structure fires.[6]

◆ MODEL ARSON LAWS

Arson is best discussed in terms of the laws under which the crime is investigated. The model arson law as developed by the International Fire Marshals Association is a part of the laws of many states.[7] This law is divided into four sections as to degree.

First-degree arson under the model law is the burning of a dwelling. Under this category, any person who willfully and maliciously sets fire to, burns or causes to be burned, or aids, counsels, or procures the burning of a dwelling is guilty of arson. The model law implies that the offense is the same whether the building is occupied, unoccupied, or vacant. It further indicates that the section applies whether the property is the property of the accused or of some other person. The section includes structures that are parts of the dwelling, such as kitchens or shops. The sentence provided under this section is generally from 1 or 2 years to 20 years.

Second-degree arson under the model law is burning buildings other than dwellings. This section also requires that the fire has been willfully and maliciously set, and it provides that an individual is chargeable who has either set the fire, caused it to be burned, or has aided, counseled, or procured the burning of the building. Again, the offense is applicable whether the property belongs to the person charged or to another. The statute often provides for a possible penalty of 1 to 10 years.

Third-degree arson in the model law pertains to the burning of other property. Individuals who willfully and maliciously set fire to, burn or cause to be burned, or aid, counsel, or procure the burning of any property of any class, as long as the property has a value of $25 or more and is the property of another, are guilty and may be sentenced for a period of 1 to 3 years.

Fourth-degree arson in the model law relates to attempts to burn. This provides that a willfully and maliciously set fire may be chargeable as an attempt even if combustible and/or flammable materials have been only distributed. This includes any acts preliminary to the setting of the fire. The suggested law also includes definitions to clarify the recognition of such a crime.

A separate offense category is included for the crime of burning with intent to defraud an insurer. This likewise requires that the act be willful and malicious and that the property be insured at the time of the fire.

It should be noted that the phrase *willfully and maliciously* occurs time after time in the model arson law. This means that a fire set accidentally is not chargeable under the model arson law. To obtain a conviction of arson, a prosecuting attorney must be able to prove in court that the act was willfully and maliciously perpetrated. In fact, the law enforcement officer should not consider making an arrest on an arson charge if such willful and malicious action cannot be proved. A distinction is made in the law between dwelling places and other occupancies. This is because there is a chance of death in the case of a fire that has been maliciously set in a dwelling.

A number of states have adopted the Model Penal Code of the American Law Institute as first published in 1962.[8] This code covers the entire criminal law field and encompasses some concepts that are at variance with more traditional modes.

In the Model Penal Code, arson is covered in the following manner, as excerpted from Arkansas Criminal Code:

5-38-301.—ARSON

a. A person commits arson if he starts a fire or causes an explosion with the purpose of destroying or otherwise damaging:

   1. An occupiable structure or motor vehicle that is the property of another person; or
   2. Any property, whether his own or that of another person, for the purpose of collecting any insurance therefor; or
   3. Any property, whether his own or that of another person, if the act thereby negligently creates a risk of death or serious physical injury to any person; or
   4. A vital public facility; or
   5. Any dedicated church property used as a place of worship exempt from taxes pursuant to §26-3-301; or
   6. Any public building or occupiable structure that is either owned or leased by the state or any of its political subdivisions.

b. 1. Arson is a Class A misdemeanor if the property sustains less than five hundred dollars ($500) worth of damage;
   2. Arson is a Class D felony if the property sustains at least five hundred dollars ($500) but less than twenty-five hundred dollars ($2,500) worth of damage;
   3. Arson is a Class C felony if the property sustains at least twenty-five hundred dollars ($2,500) but less than five thousand dollars ($5,000) worth of damage;
   4. Arson is a Class B felony if the property sustains at least five thousand dollars ($5,000) but less than fifteen thousand dollars ($15,000) worth of damage;
   5. Arson is a Class A felony if the property sustains at least fifteen thousand dollars ($15,000) but less than one hundred thousand dollars ($100,000) worth of damage; and
   6. Arson is a Class Y felony if the property sustains damage in an amount of at least one hundred thousand dollars ($100,000).

c. For purposes of this section, "motor vehicle" means every self-propelled device in, upon or by which any person or property is, or may be, transported or drawn upon a street or highway.

5-38-302.—RECKLESS BURNING

a. A person commits the offense of reckless burning if he purposely starts a fire or causes an explosion, whether on his own property or that of another, and thereby recklessly:

    1. Creates a substantial risk of death or serious physical injury to any person; or

    2. Destroys or causes substantial damage to an occupiable structure of another person; or

    3. Destroys or causes substantial damage to a vital public facility.

b. Reckless burning is a Class D felony.

5-38-303.—FAILURE TO CONTROL OR REPORT DANGEROUS FIRE

a. A person commits the offense of failure to control or report a dangerous fire if he knows that a fire is unattended and is endangering the life, physical safety, or a substantial amount of property of another person, and he:

    1. Fails to act in a reasonable manner to put out or control the fire when he can do so without substantial risk to himself; or

    2. Fails to act in a reasonable manner to report the fire.

b. Failure to control or report a dangerous fire is a Class B misdemeanor.[9]

Most statutes patterned after the Model Penal Code use similar language. A comparison of the Model Arson Law and the arson sections of the Model Penal Code will reveal that there are advantages and disadvantages in each.

Arkansas and all other states also have statutes relating to wildland fires. In some states a wildland arson is considered a felony; in others it is a misdemeanor.

As noted in Chapter 10, federal agencies also have statutory responsibilities in arson suppression. The Bureau of Alcohol, Tobacco and Firearms is the primary agency with this responsibility.

◆ MOTIVES FOR ARSON

Although not absolutely necessary, it is most helpful to the prosecution of a case if a motive can be established for the setting of fire. A primary motive is avarice, a great desire for wealth. In such a situation, the perpetrator feels that a settlement from the insurance company can be made on the loss without the insurer realizing that the insured has in fact set the fire to defraud the insurer. This type of arson occurs most frequently during times of poor economic conditions; however, there are exceptions to this rule.

A jury often wants to have a demonstrable *motive* considered with the evidence so that they can justify the verdict to themselves, even though it is not a legal requirement. Although motive is not essential to establish the crime of arson, and need not be demonstrated in court, the development of a motive frequently leads to the identity of the offender.

Establishing motive also provides the prosecution with a vital argument when presented to the judge and jury during trial. It is thought that the motive in an arson case often becomes the mortar that holds together the elements of the crime.

Law enforcement–oriented studies on arson motives are *offender-based;* that is, they look at the relationship between the behavioral and the crime scene characteristics of the offender as it relates to motive. One of the largest present-day offender-based studies consists of 1,016 interviews of both juveniles and adults arrested for arson and fire-related crimes, during the years 1980 through 1984, by the Prince George's County Fire Department (PGFD), Fire Investigations Division. These offenses include 504 arrests for arson, 303 for malicious false alarms, 159 for violations of bombing/explosives/fireworks laws, and 50 for miscellaneous fire-related offenses.

The study was conducted primarily because fire and law enforcement professionals were entitled to take upon themselves the task of conducting their own independent research into violent incendiary crimes. The PGFD study determined that the following motives are most often given by arrested and incarcerated arsonists:[10]

- ◆ vandalism
- ◆ excitement
- ◆ revenge
- ◆ crime concealment
- ◆ profit, and
- ◆ extremist beliefs

For the purposes of classification, FBI behavioral science research defines *motive* as an inner drive or impulse that is the cause, reason, or incentive that induces or prompts a specific behavior.[11] A motive-based method of analysis can be used to identify personal traits and characteristics exhibited by an unknown offender.[12] For legal purposes, the motive is often helpful in explaining why an offender committed his or her crime. However, motive is not normally a statutory element of a criminal offense.

The motivations discussed in this chapter are also outlined and described in the NFPA 921, *Guide for Fire and Explosion Investigations* (2001).

Fires set by individuals for political reasons are discussed in detail later in this chapter. This type of activity is not difficult to recognize.

◆ INVESTIGATION OF SUSPECTED ARSON

Successful investigation of maliciously set fires requires the establishment of a *corpus delicti*. This means that the investigator must be able to prove that a crime has in fact been committed. Every potential natural cause for such a fire must be eliminated. It will be necessary to prove beyond a reasonable doubt that the fire could not have occurred as a result of accidental or natural causes. In defending the accused, the defense attorney in an arson case will most often attempt to prove that the fire was accidental. The prosecuting attorney must make every effort to prove that the fire could not have started accidentally.

RESPONSIBILITY FOR INVESTIGATIONS

Fire investigation responsibilities are generally carried out in one of three ways:

1. Entire job, including cause determination, investigation and arrest carried out by fire service investigators.
2. Entire job, including cause determination, investigation and arrest carried out by police service investigators.

FIGURE 11.2 ◆ Fire investigation involves examination of burn pattern. *(Jackson, Michigan, Fire Department)*

3. Split responsibility, with fire service personnel being responsible for cause determination, followed by police investigation and arrest. In some communities, a team concept is employed. In others, fire service personnel merely notify the police when arson is suspected.

Many major cities, including New York, Los Angeles, Denver, Miami, and Houston, assign complete responsibility for arson suppression to the fire service. Personnel assigned must be thoroughly trained in law enforcement procedures and are empowered to make arrests. This concept has an advantage because the entire responsibility is placed on one agency, and it eliminates shifting blame for poor performance and the jealousies that may arise when two agencies are investigating the same crime. Of course, communities utilizing this concept continue to rely on police support, primarily in intelligence and patrol activities.

In some communities, police departments are vested with the entire responsibility for arson detection and investigation. This system, to be successful, entails a great deal of training for police officers in examining fire scenes. The officers must qualify as expert witnesses in examining fire scenes to obtain any convictions. A person with fire suppression experience usually qualifies much more easily.

Responsibility assigned to a joint police–fire team works well in many communities. However, it has failed in others. A problem with this concept is that work schedules in police and fire departments are generally not the same. Days off are not the same, which results in lost time when one of the team members is not on duty. If one team member goes ahead with the investigation and successfully concludes it, the other may be unhappy.

Some communities have an arrangement whereby the fire suppression officer or fire marshal, on determination of arson as the possible cause of a fire, contacts the police department. The police department assumes all responsibility for the subsequent investigation and possible arrest. Both agencies must testify in court for a successful prosecution. Press publicity often gives credit only to one of the agencies, causing resentment on the part of the unmentioned agency.

Smaller communities often depend on a state or county fire marshal's office for fire investigation services. Local fire officers or the state fire investigator may make the determination of arson. In either case the state investigator carries out the subsequent investigation and makes an arrest where appropriate. This is a desirable arrangement, especially in view of the difficulty for persons who have investigated very few fires to be qualified as experts in court. Some smaller municipalities have developed an "on-call" system for mutual aid in fire investigations in an effort to overcome the experience problem.

Mississippi law requires each of the state's 82 counties to have at least one deputy sheriff trained in arson detection and investigation. These individuals, who are trained by the state's fire academy, are able to conduct a preliminary investigation and summon a state fire investigator when appropriate.

Charlotte, North Carolina, established a unique fire investigation task force, which includes fire and police department investigators, agents from the U.S. Bureau of Alcohol, Tobacco and Firearms, and the North Carolina State Bureau of Investigation as well as a district attorney. The task force has maintained a case clearance rate above 30 percent since its formation in 1985.

State fire schools as well as the National Fire Academy provide training for fire investigators. In addition, the International Association of Arson Investigators holds instructional seminars at both the national and regional levels. Similar seminars are held in Canada.

Judges and prosecuting attorneys also need indoctrination in the arson control area. Seminars for these individuals have proven to be quite successful in creating a greater awareness of the intricacies of prosecution of arson cases. One state raised its arson conviction rates from less than 1 percent to 50 percent in less than three years by conducting seminars and mock trials for judges and attorneys.

INSURANCE FRAUD

It is desirable to obtain a full inventory of materials and contents of fire-damaged structures. This is especially important if insurers should attempt to prove that a building was overinsured. Adjusters and other people familiar with evaluation techniques can be helpful in this connection. The records of insurance carriers should be checked to determine if there has been any recent change in insurance coverage on the building. Such a change may be an indication of an attempt to defraud the insurance carrier. To prove fraud, a claim must have been filed by the insured.

Insurance companies are concerned about soaring losses as a result of arson. In an effort to combat rising losses, which are reflected in higher insurance costs for the property owner, insurance companies have intensified their arson suppression programs.

One major insurance industry project is the Property Insurance Loss Register. This industrywide program is designed to record pertinent data about fires from adjusters. Computers store names of insured tenants, owners, partners, corporate officers; types of occupancy; cause, time, and date of losses, as well as insurers and amounts of coverage. This information is available for use in investigations.

Insurance carriers deny liability on suspicious fires, which places the insured in the position of having to sue to recover alleged losses. They often employ their own investigators to assist in such cases. In many cases, the courts are upholding the denial of liability.

Some insurance companies assist communities in providing accelerant-detection canines. Companies also provide major funding for arson award programs, in which tips leading to arrests are rewarded.

THE VALUE OF CONFESSIONS

Occasionally, in the investigation of an arson case, investigators obtain a confession to the crime. Such a confession is valid if properly obtained and can be of great assistance in the successful prosecution of the case. No threats may be used nor can any promises be made in taking a statement that turns out to be a confession. It is extremely important that the individual be forewarned of full rights as required by the U.S. Supreme Court's *Miranda* decision. This decision requires that interrogation should not begin until the suspect has agreed to submit to the procedure. The individual must be warned, according to the decision, that a statement need not be made at all. The suspect should also be told that he or she may stop at any time and request the presence of an attorney; that in the event the suspect cannot afford a lawyer, one will be provided; and that any statement made could be adversely used in a court of law.

In the case of arson or malicious burning, a confession alone may be used for convicting the defendant if there is corroborative evidence connecting the individual to a fire. This means that it is not always necessary to have a witness who can place the defendant at the scene of the fire at the time of its burning. It may be sufficient to have a witness testify that the defendant was seen around the area 15 or 20 minutes before the fire or testify that the defendant was seen around the fire scene shortly after the discovery of the fire. In using such a witness, the prosecutor should consider the need to report accurate times so the defendant can be placed at or near the scene within a short time before or after the fire.

FIRE SCENE EXAMINATION

Fire scene examination is a critical part of the investigation of fires. This examination cannot take place until the fire has been brought under control; however, the scene should not be left unguarded at any time. The fireground should be securely maintained until the investigation has been completed and evidence or photographs at the scene are no longer needed.

In examining the fire scene, every effort should be made to locate possible sources of ignition, such as electrical wiring, heating appliances, carelessly discarded cigarettes, or other sources that may have accidentally caused the fire. The investigator should be alert for any possible evidence pointing to the origin of the fire. Proof must also be established that a fire did actually occur, and the investigator must have the owner or owner's agent testify as to the condition of the structure before the fire.

The firefighter plays a major part in the investigation of a maliciously set fire. Without the complete cooperation of the fire department, successful investigation and prosecution will be extremely difficult. Fire suppression officials responding to the alarm often have the first trained eyes on the scene for observing conditions. The fire department may also observe conditions relating to the spread of the fire, the use of an accelerant, and other factors.

The conditions noted in the building during fire inspections and prefire planning surveys can be most helpful in determining any significant changes in conditions at the time of the incendiary fire. The fire inspector may have recorded the original position of cabinets and other equipment that may have been moved by the person starting the fire.

On their approach, fire company officers can observe the general conditions at the scene, such as weather and road conditions. These factors may have a bearing on the subsequent investigation (through consideration of tire tracks in the snow, as an example).

The first fire department personnel to arrive should observe people and automobiles, as well as license numbers and descriptions of vehicles in the vicinity. This is not an easy task for a firefighter while fighting the fire or rescuing occupants. Any remembered detail such as a license number might be the key to tracing a suspected arsonist. A suspect can sometimes be arrested on the scene, especially because whoever started the fire is apt to stay around long enough to see that the job was successful.

The first personnel on the scene should observe the size of the fire and the speed at which it is traveling. Again, as with noting license plate numbers, this is not easy to do under the pressure of fire control and rescue problems. They should particularly observe whether separate fires may be burning, along with the intensity and rapidity of spread. This information may be extremely critical at a later time when investigation indicates that the fire was in fact maliciously set.

Other indicators that may be of value in an investigation are odors, such as those of gasoline or kerosene. Methods of extinguishing the fire might also have significance. The condition of windows and doors in the building should be observed for indications of forcible entry into the building, or signs that material has been removed from the premises. Bystanders at the fire scene may offer some clues. Familiar faces in the crowd may have been noticed at other fires or emergencies in the community.

FIRE INVESTIGATOR

When the fire is under control the fire suppression officer may decide that a fire investigator should be called to the scene. This should be done before the fire suppression forces depart. To preserve evidence, full security should be maintained at the fire scene. Police officers, security guards, or other persons having a legal responsibility may need to stand by at the fire scene while awaiting the arrival of the fire investigator. The scene should be guarded until the fire investigator has arrived.

The fire suppression officer, while awaiting the arrival of the investigator, may look for evidence of arson intent. Among things to look for are unusual odors in the building (possibly indicating the use of flammable accelerants), multiple fires, and undue charring and uneven burning. A trained investigator can readily note uneven burning and excessive or undue charring. The fire officer or investigator, or both, may look for holes that have burned through the walls or ceilings of the building. This would indicate the possible use of accelerants in the fire. There is a growing tendency for fire setters to deactivate sprinkler systems, block fire doors, and otherwise impede the operation of fire protection equipment. All of these devices should be checked to see whether they have been tampered with.

Tracks around the premises may be quite revealing. They can best be noticed by fire personnel arriving early at the fire. An observation may be made that there were people walking around the building before the fire. There is a good possibility that this will not be useful evidence later because the tracks can well be obliterated by fire service personnel carrying out fire suppression duties.

This is also true of fingerprints. Although smoke and heat may obliterate most fingerprints, occasionally some can be found by thorough processing of the area. Most fire service personnel do not have training in fingerprint techniques and may overlook the possibilities of lifting prints.

A thorough and systematic examination should be made of the fire scene to determine possible causes of the fire and to detect evidence of tampering with any equipment. Heating equipment should be checked to see if gas valves were left open; electric and gas stoves should be checked for the possibility of burners having been turned on. Electrical equipment and appliances should also be examined.

Fatal fires deserve special attention. An autopsy should be conducted in all cases to ascertain the true cause of death. Where possible the investigator should be present at the autopsy and ensure that an adequate investigation has been made including photographs of the entire body.

A search should be made to detect any possible use of flammable liquids and for containers left behind at the fire scene that may have held flammable liquids. Ignition sources to look for are matches in the area, candles, or trailers (strips of combustible materials) placed between sources of ignition. Mechanical igniters or timing devices are not frequently found but have on occasion been used to start a fire. And not to be discounted is the possibility that the fire was started as a result of glass magnifying the rays of the sun. At the same time, excessive amounts of debris, rags, waste, and accumulations of papers or other combustibles in the area should be noted. All these factors should be carefully assessed and pointed out in the fire investigator's report.

PRESERVATION OF EVIDENCE

Every effort must be made to preserve and protect any evidence of the cause of the fire that is found during the fire investigation. Extreme care is needed in using water around the fire scene where evidence may be involved. This precaution is not always easy to observe. Care should be taken in salvage and overhaul operations. There is often a conflict between investigation and suppression activities. Overzealous cleanup activities sometimes result in the removal or destruction of necessary evidence. This will improve with adequate fire suppression training as it relates to fire scene examination and investigation. The owner or occupants of the building should under no conditions be permitted to return until the investigation has been completed. Visitors, including reporters, should not be allowed to enter the fire scene.

Of major concern to fire investigators is the preservation of evidence to prevent its destruction or alteration before it is submitted for competent examination and analysis. Destruction or alteration of evidence, particularly when it will be the subject of pending or future litigation, is often referred to as *spoliation*.

The impact of failing to prevent spoliation can result in testimony being disallowed, sanctions, and potentially civil or criminal remedies.[13] Standards presently establish practices for examining and testing items of evidence that may or may not be involved in product liability litigation (American Society for Testing and Materials). In laboratory examinations where the evidence will be altered or destroyed, all persons involved in the present or potential cases should be given the opportunity to make their opinions known and be present at the testing.

The use of barricades can be helpful if other means of restraining entry are ineffective. The fire service generally has the legal means of keeping unauthorized people from entering the structure while fire fighting operations, including overhaul, are under way.

The fire officer should make a note of the time evidence is discovered. If at all possible, some other person should also witness the discoveries and verify the conditions found. This verification may be a key factor in court.

There should be a mandatory procedure for handling evidence so that it is not damaged or distorted. The evidence should be maintained at its original location until collected and entered into evidence. Prior to its removal, all evidence should be photographed in place and a sketch developed identifying its exact location in the fire scene. All evidence must be marked or labeled for identification at the time of collection. Physical evidence should be stored in a secured location designed and designated for this purpose.

When collecting physical evidence for examination and testing, it is often necessary to also collect comparison samples. This is especially important when collecting materials believed to contain liquid or solid accelerants. The laboratory can evaluate the possibility that flammable liquids were introduced to the scene and were not the normal fuels present.

All of these factors should be carefully assessed and included in the fire investigator's report to include photographs, videos, diagrams, and sketches. A recommendation for adequate documentation of a fire scene is found in NFPA 921 (*Guide for Fire and Explosives Investigation*), Chapter 8, published by the National Fire Protection Association.

CONTROL OF STATEMENTS TO THE PRESS

Statements to the press regarding fire causes should be carefully controlled. Many investigations have been jeopardized because of the unwise release of information to the news media. Some fire service personnel have a tendency to make statements to the press regarding the details of fires without having specific knowledge of conditions. It is very easy to make comments to the press that later turn out to be incorrect. The most desirable approach is to have one individual responsible for all press releases. Statements about fire causes made by authorities, including elected officials, who are not members of the fire service may be most damaging to the prosecution should an arson case be developed.

◆ ARSON AND CIVIL UNREST

Multiple arsons during periods of civil unrest during the 1960s and 1970s brought back an arson motive that had not been prevalent for many years. Such fires impose a severe strain on routine investigative procedures.

Tulsa, Oklahoma, and several other cities experienced disorders in the 1920s. At least 27 fatalities occurred in the 1921 racial disturbances in Tulsa. More than 1,000 buildings, primarily residences, were destroyed by fire. Detroit experienced riots during World War II.

In 1992, Los Angeles experienced the most expansive, violent, and costly epidemic of urban unrest in the 20th century. A total of 53 civilians were killed and 2,328 were injured, including 59 firefighters. Over 850 buildings were damaged or destroyed by fire, with a total loss of over $560 million. The Fire Sprinkler Association of Southern California estimated that nearly 2.5 million square feet of building floor area were saved by action of automatic sprinklers.[14]

Operations during times of civil unrest or disorder require planning. Fire investigative personnel cannot merely sit and wait for calls and expect to have success once multiple fires start occurring. For example, it could be very useful to assign fire investigation personnel to patrol duties in the time of disorders. These patrols could detect fires early and thereby increase chances for successful suppression; they could also aid in apprehending those responsible for setting the fire. The use of unmarked vehicles is probably most desirable for this purpose. In a few cities, patrols of this kind have been equipped with portable fire extinguishers to enable them to control incipient fires. The possibility of such personnel being lured into an area for physical attack must be kept in mind.

Fires that occur during periods of disorder should be as thoroughly investigated as those that occur during normal times. During periods of unrest, many people set fires for the first time. It is easy to lay aside investigation of minor fires during a disorder in deference to fires having a great monetary loss or loss of life. Failure to investigate these smaller fires may serve only to improve the confidence of their setters and to allay their fear of being apprehended.

During periods of unrest there may be, in addition to a multitude of set fires, a number of false alarms and bomb threats. These calls must receive attention because they could be genuine. Individuals have not hesitated to use explosive devices during periods of disorder. False alarms may be intended to send fire equipment to the other side of town to permit the unabated spread of planned fires.

Fires must be investigated in such a manner as to provide adequate information in the event an arrest is made. The possibility of a property owner deliberately setting fire to the premises during a period of unrest cannot be overlooked. Fires occurring during civil disorder may encourage individuals who are naturally inclined to set fires. An impulsive fire setter, for example, could be strongly influenced by television publicity given to fires during disorders.

Fire investigation during a civil disturbance involves a number of unusual problems. Guarding the scene of the fire for evidence preservation may be extremely difficult. Angry mobs may make it impossible to maintain effective safeguards. Also, the same building may be set on fire several times during the course of the night. Procedures normally followed in preserving evidence must be laid aside. Members of the National Guard may be asked to guard the fire scene until the investigator can safely return.

Some communities have investigators respond to fire scenes along with or just behind fire apparatus during periods of disorder. This gives them an opportunity to locate and photograph evidence before it is removed. Fires set during disorders are usually rather crudely set.

Collecting and preserving evidence is just as important for fires set during times of civil disorder as for any other type of maliciously set fire. There is no reason fires cannot start from natural or accidental acts during times of disorder as well as during less turbulent times. The case will come to trial under quieter conditions. The investigator must be prepared to testify as to whether the fire could have been of natural or accidental origin. At least, it must be shown that an examination of heating appliances, electrical systems, stoves, flammable-liquid operations, etc., did not indicate these to be the cause of the fire.

If flammable liquids were used, the investigator should determine, if possible, where the flammable liquids were purchased and the means used to transfer them to the premises. Containers are, of course, necessary to transport flammable liquids.

When many fires occur in a short period of time, handling evidence generally becomes an issue. It is imperative that the evidence be adequately marked and properly stored. Materials gathered at the fire may present a problem as portions of fire bombs are often quite fragile. A scoop-type box is useful for this purpose.

◆ **ARSON ARRESTS**

Many law enforcement and fire service investigators consider arson to be a difficult crime to solve with an arrest. While arson convictions may require diligent efforts, the *Uniform Crime Reports* indicates that in 2002, law enforcement agencies in the nation cleared 16.5 percent of arson offenses by either arrest or exceptional means. Cities collectively cleared 15.9 percent of arsons while rural counties cleared 23.0 percent of arson offenses, and suburban counties cleared 17.1 percent. The Northeast geographical region cleared 19.8 percent of the arson incidents; the South, 18.9 percent; the Midwest, 14.8 percent; and the West, 13.9 percent.

Forty-three percent of arsons cleared in the nation in 2002 involved juvenile offenders (those aged under 18). In cities, 45.6 percent of arsons cleared by arrest involved juveniles. Community/public structures represented 71.8 percent of the clearances involving juveniles, and motor vehicle 21.9 percent.[15]

JUVENILE FIRE SETTERS

As noted in the foregoing statistics, the juvenile fire setter is a major problem in the arson field today. The term *juvenile fire setter* includes both those children under the age of responsibility as well as those who can legally be held responsible for their acts. Some jurisdictions use the term *child fire setter* because the word *juvenile* may be associated with *juvenile delinquents,* causing parents to be reluctant to report their child's fire-setting activities to authorities.

Studies have indicated that a high percentage of fires are set by juveniles, many by boys under the age of 10, although fire setting by girls is increasing. A number of model programs are in use to address this problem. Most involve long- and short-term counseling, fire safety skill instruction, fire prevention projects, and parental awareness, as well as extensive networking.

Juvenile court, the school system, state family and children agencies, and a wide variety of other human service agencies and organizations are often involved in the programs in addition to the fire department. The turning in of false alarms and damaging of fire safety equipment are also addressed by these cooperating groups. The age of the involved child is, of course, a major consideration in determining the proper remedial process.

COURT DECISIONS

Two U.S. Supreme Court decisions have a bearing on the investigation of fires. The 1978 decision, *Michigan v. Tyler,* established guidelines for right of entry to investigate cause and origin and to subsequently gather evidence should the fire be determined as incendiary.[16] The Court held that the original entry for fire suppression and subsequent investigation to determine cause and origin is permissible without a warrant; however, subsequent entries may necessitate obtaining consent to enter or a warrant.

In a later case, *Michigan v. Clifford,* the Court reaffirmed its findings in the earlier case and recognized the ability of the investigator to seize evidence in clear view at time of entry.[17] Since both cases directly relate to the investigation of fires, details of the decisions should be studied by fire investigators.

Following the lead of the U.S. Supreme Court, state courts have indicated similar restrictions. In a South Dakota Supreme Court case, *State of South Dakota v. Jorgensen,* the court upheld use of evidence found during the state fire investigator's initial entry while the fire was still smoldering but upheld suppression of evidence found in later, warrantless entries.[18]

The recent trend is that U.S. Supreme Court decisions now, more than ever, continue to define the admissibility of expert scientific and technical opinions, particularly as they relate to fire scene investigations. These decisions impact how expert testimony is accepted and interpreted.

A judge has the discretion to exclude testimony that is speculative or based upon unreliable information. In the case, *Daubert v. Merrell Dow Pharmaceuticals,* the Court placed the responsibility upon a trial judge to ensure that expert testimony was not only relevant but also reliable.[19] The judge's role is to serve as a "gate-keeper" to determine the reliability of a particular scientific theory or technology. The Court defined four criteria to be used by the gatekeeper to determine whether the expert's theory or underlying technology should be admitted. *Daubert* allows the Court to gauge whether the expert testimony aligns with the facts of the case as presented.

◆ ARSON-RELATED RESEARCH PROJECTS

The National Center for the Analysis of Violent Crime, located at the FBI Academy in Quantico, Virginia, conducted a comprehensive study of serial arsonists. *Serial arson* is defined as an offense committed by fire setters who set three or more fires with a significant cooling off period between the fires.

Researchers reviewed records of almost 1,000 incarcerated arsonists and in turn conducted detailed interviews of 83 confirmed serial arsonists in their places of confinement. The research was reported in *A Report of Essential Findings from a Study of Serial Arsonists.*

In addition to the sex and racial breakdown revealed by arrest records, the study indicated that almost half of the serial arsonists interviewed had some type of tattoo, while almost one-fourth had some form of physical disfigurement. Over 60 percent had multiple felony arrest records for offenses other than arson, and almost two-thirds had multiple misdemeanor arrest records. Over half had been held in juvenile detention in earlier years. One-fourth reported at least one suicide attempt.

A majority of the serial arsonists grew up in middle-class neighborhoods. Relationships with mothers were much closer than with fathers in a high percentage of the cases. Many started setting fires at very young ages. The 83 offenders interviewed had set a total of 2,611 fires, an average of 31.5 arsons per offender.

Of special interest to investigators is the fact that offenders reported they had been questioned by law enforcement officers an average of 3.7 times each before being arrested. The arsonists indicated that they had set an average of 25.3 fires without being questioned. They indicated that efforts by law enforcement investigators were responsible for 38 percent of the apprehensions. In 21 percent of the cases, the individual went to law enforcement agencies and confessed, while informants were responsible in 7 percent of the cases. Witnesses were the key element in 12 percent of the arrests.

Accomplices assisted in 20 percent of the fires. In a majority of cases (59 percent), material available at the scene was the accelerant of choice. Gasoline and other petroleum products were used in 28 percent of the fires. Matches were the predominant ignition device.

Thirty-one percent of the serial arsonists remained at the scene of the fire; 28 percent went to another location. Slightly over half returned to the scene at some time after the fire, usually within 24 hours of the arson.

Nearly half of the serial arsonists consumed alcohol before setting fires. Over half cited revenge as the motive for their arsons. The next most prevalent motive was that of "excitement." Nearly one-half entertained no thought of ever being arrested for their crimes. Almost one-half admitted that their acts were premeditated as opposed to impulsive.

A review of the findings indicates that serial arsonists generally lack the skills to deal with problems of life in general. They were failures in interpersonal relationships and in occupational activities. The report states that "For many, arson may be the only thing they have tried in their life that yields relative success."[20]

The Bureau of Alcohol, Tobacco and Firearms, U.S. Department of the Treasury, in 1992 to 1993 sponsored a research project on the relationship between drug activity and arson. The U.S. Fire Administration, the International Association of Chiefs of Police, and agencies having fire investigation responsibilities in five cities also participated. The project was conducted by the Virginia Commonwealth University in Richmond.

The study identified the following problem areas in arson detection, investigation, and reporting:

(1) Unique characteristics of arson impede detection, investigation and reporting; **(2)** There is a lack of agency participation in reporting arson; **(3)** Lack of uniformity in measuring and defining arson variables impedes accurate statistical accounting of arsons; **(4)** Jurisdictional and organizational problems confound arson investigations; and **(5)** There is an immediate need for standardized and comprehensive arson investigation training.

The percentage of drug-related arsons in the five cities studied ranged from a high of 24 percent in Kansas City, Missouri, to a low of 13 percent in New Haven, Connecticut. Arsons represented 34 percent of all fires in both cities during the study period. The five-city average for drug-related arsons is 22 percent, thereby indicating that the drug–arson association is a definite factor in the arson-suppression picture.

The study recommends additional research on drug-related arsonists, which would in turn enhance prevention strategies.[21]

Another study illustrated the value of thorough fire investigations followed by vigorous prosecution. In this project, which was prepared for the Law Enforcement Assistance Administration, arson incidents, arrests, and convictions were analyzed in 108 cities over a four-year period. Cities ranking in the upper third according to arson arrest rates had 22 percent less arson per 100,000 population than cities ranking in the bottom third; cities in the upper third, according to conviction rate, had 26 percent less arson.[22]

◆ ARSON CONTROL NEEDS

The International Association of Arson Investigators and the National Fire Protection Association sponsored a symposium of Arson Control in the 1990s in Houston. Nationally recognized speakers from law enforcement, fire services, the insurance industry, and legislatures participated in this program. Several of their conclusions follow:

- Automatic sprinklers, public fire safety education, smoke detectors, and code enforcement all play an important role in arson suppression.
- Training for fire investigators must be improved. In some cases, the arsonist is more sophisticated than the fire investigators.
- Safety and health issues for fire investigators need additional study.
- Key members of state legislatures need to be made aware of arson problems and solutions.
- Public awareness of the arson problem is essential to effective abatement. Accurate data are of fundamental value in this effort.
- Wildland arson problems are becoming closely associated with suburban and urban areas as the trend toward construction of homes in woodlands continues.
- Some aspects of programs that address serial arsonists and those that relate to juvenile fire setters are similar. Interconnection responsibilities should be explored.
- The sheer volume of motor vehicle fires creates a gap in terms of thorough investigation. This gap needs to be addressed.[23]

In 1996, the Bureau of Alcohol, Tobacco and Firearms received funding from Congress to construct a Fire Investigation, Research, and Education (F.I.R.E.) facility to assist investigators in understanding and reconstructing the physical effects of fire in the buildings. The F.I.R.E. Center has become an internationally recognized laboratory that provides law enforcement agencies, and other fire investigators across the nation, access to a source of scientific research and forensic support needed to determine the causes and characteristics of suspicious fires. It has also become a repository for the collection of scientific facts, experimental results, material property data, and other knowledge related to fire incident investigation, analysis, and reconstruction research.

FIGURE 11.3 ◆ City buses can be used to spread a message in fire/arson prevention.

The suppression of arson through the detection, investigation, and control of maliciously set fires has an important place in efforts directed toward prevention. Arson remains a serious problem and is a significant cause of fires in this country. A coordinated effort between police and fire agencies in the investigation of suspicious fires to bring about prosecution of arsonists is needed in many communities. There is a need for more qualified fire investigators in the field and thorough training in investigative techniques to discover and preserve evidence of arson that will hold up in the courts. The crime of arson often goes unrecognized because of the lack of thorough investigation or because of failure to develop sufficient information to determine that arson has occurred.

The model arson law, developed by the International Fire Marshals Association and adopted in the laws of many states, defines four degrees of arson with corresponding penalties. The phrase *willfully and maliciously* is the key factor in these provisions. To obtain a conviction of arson, a prosecuting attorney must be able to prove in court that the fire was in fact willfully and maliciously perpetrated. Some states have adopted the Model Penal Code of the American Law Institute, which differs in some respects from the model arson law.

It is helpful to the prosecution of a case if a motive can be established. Motives for arson range from insurance fraud, revenge, and cover-up of another crime to satisfaction of the abnormal urges of the vanity fire setter and the pathological fire setter.

A valid approach to the suppression of arson is a thorough investigation of all fires. No longer can the fire officer assume that a fire is accidental in origin just because the location or the owner is well-known in the community. Because motives for a deliberately set fire may not be readily apparent, no presumptions can safely be made.

A systematic examination of the fire scene is all-important in establishing the origin of the fire, whether accidental or not. The fire service personnel who arrive first should observe general conditions: people and vehicles in the vicinity, the presence of distinctive odors, the intensity and travel speed of the blaze—all of which can be clues for the investigation. If arson is suspected, a fire investigator is usually called to the scene. A trained investigator will look for any traces of incendiary origin of the fire, such as the use of flammable liquids or accelerants and evidence of tampering with equipment, such as gas valves left open on heating equipment or stove burners turned on. Photographs may also be taken at the scene. All evidence should be delineated in the fire investigator's report, and all physical evidence should be carefully preserved for laboratory examination and possible use in criminal court proceedings.

Arson is especially prevalent during periods of civil and economic unrest. Difficulties of investigation, collection, and preservation of evidence are compounded in fires that occur during these periods because of multiple fire calls, false alarms, and confusion and mob intervention at the scene.

Statistics of the FBI *Uniform Crime Reports* show arson and malicious burning as a major crime problem in the country, and figures prepared by the National Fire Protection Association show that losses from incendiary and suspicious fires have increased alarmingly in the last two decades. This is a serious matter for the fire preventionist. The only way to combat it is to persist in efforts toward exposing and suppressing the crime of arson.

■■■

Review Questions

1. The first arson law in Maryland was enacted in
   a. 1638
   b. 1697
   c. 1742
   d. 1776
   e. 1803
2. According to the NFPA, arson showed a decrease of _____ from 1978 to 2001.
   a. 21 percent
   b. 31 percent
   c. 41 percent
   d. 51 percent
   e. 61 percent
3. In 2002, of residential occupancies, single occupancies comprised approximately _____ of the total residential arsons.
   a. 50 percent
   b. 60 percent
   c. 70 percent
   d. 80 percent
   e. 90 percent
4. The Model Penal Code of the American Law Institute was first published in
   a. 1953
   b. 1955
   c. 1958
   d. 1962
   e. 1967
5. Successful prosecution of arson required the establishment of
   a. motive
   b. *corpus delicti*
   c. habeus corpus
   d. tort
   e. juris prudence
6. Motivations for arson are described in NFPA
   a. 1001

b. 1500
c. 1710
d. 921
e. 231
7. Fire investigators' responsibilities can be assigned to
   a. fire departments only
   b. police departments only
   c. shared fire and police departments
   d. all of the above
   e. none of the above
8. Possible sources of ignition include
   a. electrical wiring
   b. heating appliances
   c. cigarettes
   d. all of the above
   e. none of the above
9. The 1992 riots in Los Angeles killed _____ civilians.
   a. 15
   b. 26
   c. 39
   d. 47
   e. 53
10. Approximately _____ of arson offenses are cleared by arrest or exceptional means.
    a. 6 percent
    b. 11 percent
    c. 16 percent
    d. 21 percent
    e. 25 percent
11. The 1978 decision *Michigan v. Tyler*
    a. established guidelines for entry to investigate cause and origins
    b. allowed expert testimony for arson
    c. set criminal sentencing guidelines
    d. all of the above
    c. none of the above

■■■

Answers

1. a
2. e
3. c
4. d
5. b
6. d

7. d
8. d
9. e
10. c
11. a

Notes

1. *Assembly Proceedings,* February/March 1638/39 (St. Mary's City, Md.), p. 72.

2. Betty Richardson and Dennis Henson, *Serving Together: 150 Years of Firefighting in Madison County, Illinois* (Collinsville, Ill.: Madison County Firemen's Association, 1984).

3. Steven R. Frady, *Red Shirts and Leather Helmets* (Reno, Nev.: University of Nevada Press, 1984), pp. 144–145.

4. U.S. Department of Justice, Federal Bureau of Investigation, *Crime in the United States: Uniform Crime Reports, 2002* (Washington, D.C.).

5. *FBI Crime in United States,* Section II, pp. 41–43.

6. National Fire Protection Association Report on U.S. Arson Trends and Patterns (Quincy, Mass., 2002).

7. *Model Arson Law* (Quincy, Mass.: National Fire Protection Association, 1931).

8. *Model Penal Code* (Philadelphia: American Law Institute, 1962), p. 152.

9. Arkansas Code of 1987, Annotated, as amended 1997.

10. D. J. Icove and M. H. Estepp, "Motive-Based Offender Profiles of Arson and Fire-Related Crimes," *FBI Law Enforcement Bulletin* (April 1987).

11. A. O. Rider, "The Firesetter: A Psychological profile." *FBI Law Enforcement Bulletin* (June/August 1992).

12. D. J. Icove, J. E. Douglas, G. Gary, T. G. Huff, and P. A. Smerick, "Arson" (Chapter 2), in J. E. Douglas., A. W. Burgess, A. G. Burgess, and R. K. Ressler, *Crime Classification Manual* (New York: Macmillan, 1992), pp. 165–166.

13. G. E. Burnette, "Spoliation of Evidence: A Fire Scene Dilemma," *Interfire,* www.interfire.org.

14. *Los Angeles Civil Disturbance, April 29, 1992* (Los Angeles Fire Department, 1992), pp. 227, 229, 248, 250.

15. U.S. Department of Justice.

16. *Michigan v. Tyler,* 436 U.S. 499, U.S. Supreme Court, 1978.

17. *Michigan v. Clifford,* 52 U.S. 4056, U.S. Supreme Court, 1984.

18. *State of South Dakota v. Jorgensen,* Opinion No. 13815, South Dakota Supreme Court, 1983.

19. *Daubert v. Merrell Dow Pharmaceuticals, Inc.,* 509 U.S. 579 1993.

20. Allen D. Sapp, Timothy G. Huff, Gordon P. Gary, David J. Icove, and Philip Horbert, *A Report of Essential Findings from A Study of Serial Arsonists* (Quantico, Va.: National Center for the Analysis of Violent Crime, 1994).

21. *Drug Related Fires in the United States* (Richmond: Virginia Commonwealth University, 1993).

22. Aerospace Corp. Survey and Assessment of Arson and Arson Investigation prepared for Law Enforcement Assistance Administration (El Segundo, Calif., 1976).

23. *Report on Arson Control in the 1990s: A Symposium* (Houston, Tex.: International Association of Arson Investigators and National Fire Protection Association, 1991).

CHAPTER 12 International Practices in Fire Prevention

In any attempt to assess current fire prevention programs on this continent, it is desirable to compare our procedures with those in effect in other parts of the world. Some fire prevention practices followed in the United States and Canada can be linked to procedures in other countries of the world; however, there are probably as many differences as there are similarities.

Assumptions may be made that all countries follow the same fire prevention practices. This is not true. Procedures practiced in Italy, for example, are quite different from those practiced in Hong Kong. The fire preventionist can learn a great deal from a study of these various procedures.

In making comparisons among countries, many other factors must be taken into account in analyzing fire prevention practices and their results. Standards of living, industrialization, customs, dollar valuations, use of electricity, even smoking habits vary considerably from country to country. All of these factors, as well as a number of others, may influence the fire situation in a given country.

An intangible element that makes a great difference is the general degree of adherence to constituted authority. A tightly controlled population where the day-to-day lives of citizens are closely watched by the government produces an entirely different fire potential than in a country where the citizens are free to act without close control. The availability of insurance is a factor; in some countries it is difficult to obtain. Travel restrictions within a country may also have a bearing on the fire situation. In some countries, travel from one section to another is severely restricted, which limits the possibility of an individual setting a fire in one community and escaping to another.

Heating methods employed are a major factor. In some countries, very few open-flame devices are available or used. In others, open-flame devices are quite common. There are great varieties in length of heating seasons. Practices in construction likewise vary considerably from place to place. In some countries, practically all construction is masonry or fire-resistive. In others, the common practice is to use wood, or even cardboard, for most buildings.

Fire prevention practices in a country where construction is primarily noncombustible are not the same as in those where construction is primarily wood frame or plywood. Fire deaths, however, may not be as directly related to construction as is commonly thought.

At any rate, comparisons from country to country must be made with the previously mentioned factors in mind. Conclusions made without consideration of these factors may be invalid.

◆ FIRE DEATH STATISTICS FOR VARIOUS COUNTRIES

Before making a detailed study of fire safety procedures in other countries, it is desirable to review fire loss statistics from those nations. Again, in reviewing fire loss statistics, differences in valuations, standards of living, etc., must be considered. As an example of the country-by-country comparison of fire occurrences and losses, a comparison may be made between the United States and several other countries. It should be noted that procedures, problems, and fire loss statistics are generally similar in the United States and Canada.

Fire death statistics show the United States with a high death rate: 15.5 fire deaths per one million population. Roughly two-thirds of those deaths occurred in a victim's own home.[1]

The United Kingdom's fire death rate is 11.1 per million. However, there is an increase to the next highest industrialized nation, Canada, with 12.2 fire deaths

FIGURE 12.1 ◆ Fire Deaths

| Country | Adjusted Figures (Fire Deaths) | | |
	1998	1999	2000
Singapore	11	2	7
Switzerland	41	40	
Spain	250	275	260
Australia	140	140	125
Italy	435	420	410
Germany	650	630	585
Slovenia	22	17	15
France	580	575	
Czech Republic	100	110	
New Zealand	53	37	33
U.K.	690	655	645
Canada	370	425	380
Norway	53	61	57
Greece	145	120	
Poland	505	560	515
Austria	55	53	225
Sweden	180	115	110
USA	4,400	3,900	4,400
Denmark	79	84	87
Japan	2,100	2,150	2,050
Finland	91	105	94
Hungary	205	190	200
Ireland	61	85	

(continued)

FIGURE 12.1 ◆ (continued)

Population Comparisons for Fire Deaths

Country	Deaths per 100,000 persons
Singapore	0.16
Switzerland	0.64 (1997–1999)
Spain	0.66
Netherlands	0.68 (1994–1996)
Australia	0.71
Italy	0.73
Germany	0.76
Slovenia	0.91
France	0.95 (1997–1999)
Czech Republic	1.02
New Zealand	1.08
U.K.	1.11
Canada	1.26
Norway	1.28
Greece	1.34 (1997–1999)
Belgium	1.35 (1995–1997)
Poland	1.36 (1999–2000)
Austria	1.37
Sweden	1.53
USA	1.55
Denmark	1.57
Japan	1.66
Finland	1.87
Hungary	1.96
Ireland	1.97 (1997–1999)

Note: Population figures used are derived from the United Nations
Demographic Yearbook.
Source: The Geneva Association, *World Fire Statistics No. 19,* Tables 3 and 4
(October 2003).

per million.[2] The Netherlands has 9.4 per million; Czech Republic has 10.2 per million.[3]

On a per capita basis, fire losses in Canada are very similar to those in the United States. Residential fires account for 85 percent of fatalities and 46 percent of dollar loss. The latter figure is slightly higher than that for the United States. Canada had a comprehensive nationwide fire reporting system for most of the 20th century, and comparative statistics are quite meaningful. Causes of fires are likewise fairly similar to those south of the border.

Russia has 18,000 fire fatalities a year, nearly five times the rate in the United States, according to an article describing a 2003 dormitory fire in Moscow that took 32 lives and injured 139. The fire occurred in a five-story university student housing structure. Newly arrived students, primarily from Africa, Asia, and Latin America, were housed in the building while in quarantine awaiting medical checks before starting classes. Many jumped from upper stories, according to the article.

The article notes that the Russian fire fatality rate is nearly five times that of the United States with half the population. Britain's fire fatality rate is described as one per 100,000 population compared to Russia's 12.5 per 100,000. Fire experts attribute Russia's rise in fatalities to lower public vigilance and disregard for safety standards.[4]

COMPARISON OF NUMBER OF FIRES

In the United States, 2.5 fires per 1,000 people occur in buildings.[5] This compares with 0.9 per 1,000 in The Netherlands, and 0.9 in the United Kingdom.[6] Per capita fire losses also vary considerably from nation to nation; however, the United States and Canada remain near or at the top in most compilations.

These differences have existed for many years. Mr. C. A. Hexamer, President of the National Fire Protection Association in 1906, observed that in the previous year per capita fire loss in the United States had been $2.47, although the average per capita fire loss for six European countries for the same year had been $.63.[7]

Mr. Franklin H. Wentworth, Secretary, National Fire Protection Association, addressed the comparison with Europe in a series of talks in 1910. His comments included the following point:

> The European peoples point to the fire waste of America as evidence that the American people are the most careless and irresponsible individuals in the civilized world. This is not a flattering tribute. It behooves us to inquire if the charge be true, and if so, how we may mend our unpleasant reputation. Statistics show that in the United States and Canada the fire waste is roughly ten times as much per person as in Europe. This contrast is partly explained by the facts that there are more people in Europe upon whom to figure this percentage and that more buildings in America are constructed of wood. Outside of Constantinople with its unsanitary conditions, European cities are seldom visited by such sweeping fires as have devastated the American cities of Chicago, Boston, San Francisco, Atlanta, Baltimore, Chelsea, Salem, Nashville, Augusta, and others. American building construction is inferior to that of Europe in other respects. A poorly built city with numerous wooden buildings awaits only the right kind of fire on the right kind of night for its complete destruction.[8]

This condition also prevailed in 1930, as noted by Henry G. Hodges in *City Management*. Hodges stated, "The 1930 fire losses in the United States were greater than in all Europe combined. The number of fires per thousand of population in the average American city is six times as great as it is in European cities. With approximately the same population, Berlin's fire loss is one-tenth that of Chicago." Even then U.S. losses per fire, however, were $241 as compared with European losses of $668 per fire.[9]

COMPARISON OF LOSSES PER FIRE

It should be noted that the monetary loss per fire is generally greater in other countries than in the United States and Canada. This is probably attributable to several factors. One of these is the fact that North American citizens are probably less reluctant to summon the fire department than is the case in some other parts of the world. There is no penalty imposed on an individual who calls the fire department in the United States and Canada unless it can be proven that it was a malicious false alarm.

In the United States and Canada, fire stations are more closely spaced than in a number of other countries of the world, thus fire apparatus is more likely to arrive in a shorter period of time. In a number of countries, telephones are few and far between, which makes it difficult for a person to transmit a fire alarm. There are fewer fire extinguishing systems in most other countries than in the United States and Canada.

◆ CAUSES OF FIRES

A comparison of statistics on causes of fires can also be interesting. Variables in assigning fire causes, however, makes these figures less valuable than the previously mentioned comparisons because of classification. For example, reports of the London Fire Brigade[10] include as a cause of fire "structural defects" and "fires in course of business operation." Neither of these normally appears as a cause in the United States or Canada. They are contributing factors, possibly, but are not actually specifically thought of as causes of fires.

Primary causes of fires in Moscow are careless adults, careless children, heating installations and irons, followed by lightning and "others." In St. Petersburg, 45 percent of fires are caused by general carelessness, 18 percent by lightning, 13 percent by domestic electricity, and 12 percent by chimneys and stoves. Careless children cause 7 percent. An estimated 46 percent of St. Petersburg fires occur in domestic buildings, and 20 percent are in offices.[11]

A review of the Italian reports again shows the problem with attempting comparisons of statistics. Causes as listed by the Italian Fire Service, which incidentally is a nationalized fire service, include "heating, cooking, chimney, flues, ashes"; "electrical"; "unknown"; and "miscellaneous known causes." This grouping is quite different from that employed in many other countries of the world. The "miscellaneous known causes" classification could include causes that are specifically reported in other jurisdictions. It is difficult to make meaningful comparisons between these figures and those developed in the United States and Canada.[12]

In 1999 slightly over 400,000 intentional fires were reported to U.S. fire departments, the lowest number since comparable statistics began in 1980. And 600 civilians died in these fires, again the lowest number since 1980. In Canada there has been no consistent trend up or down in total set fires. In the United Kingdom there has been a consistent and substantial trend upward, especially for vehicles. Japan has also experienced a consistent upward trend.[13]

◆ RURAL FIRES

All figures quoted so far for the various countries include rural areas as well as urban areas. The variables that arise through inclusion of rural fires again color any comparisons that may be made. There is a wide difference among countries in providing fire service in rural areas. Some countries have practically no organized fire departments in rural areas; others have fire departments in practically every section. The existence of a department would affect the loss-per-fire figure, as well as the total fire loss figure. A 100 percent relationship cannot be assumed, however, because fires in areas without fire departments may in some cases not be reported. Without benefit of a fire reporting procedure, no one in an official capacity may be aware of the existence of burned-down buildings.

Variables from country to country, and even within a particular country, make it difficult to assess comparative fire loss figures. Following are some of the allowances that must be made in evaluating statistics.

PLACEMENT OF FIRE STATIONS AND EQUIPMENT

In reviewing the total fire loss figures and the loss-per-fire figures, the placement of fire stations has been mentioned as a factor. Countries in Europe as well as in Asia generally have very few fire stations considering the size of the city; cities in the United States and Canada maintain many more.

Equipment used in the control of fires is another consideration. European fire stations, especially those on the continent, are provided with a much greater number of units per station. It is not unusual for a continental fire station to have six or seven pieces of apparatus operating out of the same station. This arrangement is seldom found in U.S. and Canadian fire stations. The single-engine substation found in many North American cities is less common in Europe.

Although the European system involves far fewer stations, a larger number of active fire fighting companies are found in each station. The arrival of these units at the fire scene at approximately the same time creates a different tactical situation than does the American system of having units arrive on the fire scene on a staggered basis, depending on the distance from the station. The American system usually provides for earlier application of water on the fire; however, difficulties may be encountered in getting major operations under way.

This has an affect on loss experience as well as on fire prevention practices. The neighborhood fire station concept should bring about a greater awareness of fire safety in a neighborhood. It stands to reason that more positive contact with the public can be made through the latter concept. Personnel assigned to a neighborhood station are aware of fire problems in the neighborhood. Members of a more remotely stationed company may not be able to develop this personal contact with citizens of the community they serve. Even in some heavily built neighborhoods, the fire station is looked on as an integral part of the community.

CONSTRUCTION PRACTICES

As previously mentioned, construction practices vary from continent to continent and even within countries, states, and provinces. However, a higher percentage of fire-resistive and noncombustible construction is employed in Europe than in the United States and Canada. European construction practices prevail in Central and South America. Even though a majority of European structures, including those for residential occupancy, are primarily of noncombustible outer wall construction, many have wood frame interior construction. This construction is intended to prevent conflagrations. The tenant, owner, or occupant will have a personal interest in fire safety, thereby mitigating results of fires from within. Experiences of the European countries during World War II further emphasized the need for fire resistive outer wall construction. Conflagration potential is also considered.

The United States and Canada have a high percentage of combustible construction. This type of construction has been prevalent throughout the history of both countries and has been responsible for many major conflagrations. Included are many fatal fires in which construction was a factor.

Availability of materials is a factor in construction practices. In the United States and Canada, building methods also vary according to availability of materials within given areas. As an example, the ready availability of wood in the Pacific Northwest states has resulted in a higher percentage of wood frame construction than is found in eastern cities.

OCCUPANCY FACTORS

A study of housing ownership may shed some light on fire experience. A fairly high percentage of people living in European countries die in the same house in which they were born. It is not uncommon for citizens of most countries of the continent to live their entire lives in the same structure.

The average American or Canadian, on the other hand, moves from dwelling to dwelling several times during a lifetime; studies indicate an average of seven moves, in fact. It is unusual to hear of people who die in the house in which they were born.

An individual who lives an entire lifetime in the same house is probably more apt to value the home and its contents than is a more transient individual. The stationary individual will probably take more adequate precautions against fire. Renters usually have less interest in home maintenance than owners.

The level of income necessary to purchase furnishings and other personal properties is less in this country and therefore decreases the significance placed on the loss of such materials in a fire.

Another factor is that of reliance on the fire department. With neighborhood fire stations, the American or Canadian may feel less resistant in summoning the fire department. Citizens consider that this is a public service to which they are entitled as taxpayers. A higher percentage of emergencies are reported to the fire department in the United States and Canada than in other countries.

On several occasions fire departments in the United States have had to make public announcements that they were unable to respond to alarms because of severe weather conditions. In recorded cases, fire alarms have dropped off to practically nothing.

In one case, in Milwaukee, emergency storm conditions lasted for over 48 hours, during which time it was practically impossible for apparatus to move out of the stations. The public reacted to radio announcements, and no serious fires occurred during the critical time. There is no way to know how long this restraint could be sustained. Indianapolis had a major flood in 1913 that lasted for four days. During this time normal water supplies were knocked out, and the fire chief announced that the fire department was helpless. No fires occurred during that time; however, they resumed when fire protection was restored.[14] Fire responses did not decrease, however, in Des Moines, Iowa, when the 1993 flood overtook its water works.

AUTHORITY IN INSPECTIONS

The matter of authority connected with inspections undoubtedly has a bearing on the occurrence of fires in a given country. A general weakness in control through inspections and other methods of enforcement can result in a greater frequency of fires than when stringent requirements are vigorously enforced.

In an exchange visit, the fire chief from St. Petersburg, Florida, visited St. Petersburg, Russia, in 1992. He found that the fire department, which is operated by the national government, has a strong fire prevention program, which includes plan reviews, issuance of certificates of occupancy, and strict code enforcement, including the ability of a fire inspector to immediately close down a business found in violation. Fire inspectors are disciplined when a preventable fire occurs in their territory. The city of almost 6 million population loses about 200 people a year in fires.[15]

◆ COMPARISONS OF FIRE PREVENTION EDUCATION AND ENFORCEMENT

Emphasis in the field of fire prevention varies from country to country. It is interesting to note that the United States and Canada apparently rely more heavily on fire prevention education than do some other countries.

In Japan, Tokyo's fire department is an example of a department with well-organized education and enforcement activities. All plans for construction must be approved by the fire department prior to the issuance of a building permit. Firefighters check structures to assure compliance during and at completion of construction.[16]

Spring and autumn fire prevention campaigns are observed each year. These include fire drills, demonstrations, and fire prevention meetings. A mobile disaster protection classroom travels over the city to educate the public in disaster protection, including fire prevention.

Fire protection equipment in buildings is inspected on a regular basis by fire service personnel. Larger buildings are required to have fire protection managers who are trained by the fire department.

Closely related to fire prevention are the public relations activities of the department. These include public fire education and arranging for fire service personnel to visit older adults who are "shut-ins" to assist them with fire precautions. Of the 18,000 members of the Tokyo fire department, 1,850 are devoted full time to fire prevention duties.

Philip Schaenman examined fire practices in Japan. He noted that many people live in wood and paper houses with very narrow gaps between houses. He states that:

> The Japanese focus on educating the public in fire-safe behavior, extinguishment of small fires by citizens using buckets of water or extinguishers, and escape practices. There is tremendous social pressure to be fire safe. Virtually the entire population is reached by local fire safety education programs starting in the home and preschool, and continuing through elementary school, junior high school, on the job, and on the street. Neighbors practice fire safety drills together. The Japanese experience demonstrates that public safety education alone can reduce fire incidence enormously.[17]

European countries strongly enforce fire safety regulations. In West Germany and other European countries, chief officers of the larger fire departments are for the most part graduate engineers. This ensures a high level of technical guidance in matters of fire suppression and gives the department a technically competent voice in the development and enforcement of fire safety standards. These chief officers do not as a rule come through the ranks of the fire brigade but enter directly into the administrative level of the fire brigade. These individuals may transfer from one jurisdiction to another, much the same as with city managers in the United States. This two-tier system is much like that of the military services in the United States and Canada.

Fire safety activities of the London Fire Brigade are described as follows:

The London Fire Brigade attaches great importance to their community programs striving to "prevent" rather than "cure" fire. Over 100,000 visits are made each year to public buildings such as theaters, offices, shops, factories, hotels, schools, health care establishments and petroleum installations. Over 15,000 sets of plans are examined as well. The London Fire Brigade also speaks to over 70,000 children between the ages of 9 and 11 each year.

Two areas of particular emphasis in the Brigade's safety campaigns are the value of installing smoke alarms and the inherent dangers of foam-filled furniture. At the urging of the country's fire brigades, Parliament implemented a restriction on the use of dangerous foam in furniture in 1989.[18]

Fire services in the United Kingdom are organized on a countywide basis except in Scotland and Northern Ireland, where a regional concept is used. Both full-time and part-time personnel are employed. This system provides full-time fire prevention personnel throughout the country, even in less populated areas protected by part-time or volunteer firefighters.

In the United Kingdom, county fire authorities have been requested by the national government to produce Integrated Risk Management Plans designed to meet five objectives including reducing the incidences of fires and reducing loss of life in fires and accidents. The thrust of this 2003 project is to require greater reliance upon locally assessed and determined standards of delivery, designed to meet local needs.[19]

Incidentally, several observers have noted that fire prevention programs in Europe are more apt to be designed to meet local needs than are those in the United States. They attribute this to the fact that those countries, despite their small geographic area, have greater success with public fire safety education than we do with a more centralized approach. Limited research appears to have been done on this conceptual difference in approach.

The provisions of the U.K. government's Integrated Risk Management Plan apply to Scotland, the section of the United Kingdom having the highest fire fatality and nonfatal injury rate. A governmental study there has recommended placing key support services, such as procurement, at the national level. It appears that the "one size fits all" is not the answer in the fire service of Scotland and probably not elsewhere as well.[20]

The National Fire Service College for the United Kingdom offers many courses relating to fire prevention. A unique program is a one-month class for architects, engineers, and building officials in which the students engage in live fire training in addition to the study of fire protection systems and related subjects. This college is considered to be the finest such facility in the world.

Construction practices favor limited fire spread in Central and South America. Climatic conditions, especially in Central America, preclude heating-system fire problems. These countries, however, are finding that fire protection and fire prevention problems are rising with the development of new technology. As an example, the popularity of liquefied petroleum gas as a fuel for heating and cooking has introduced a new fire protection problem in some of these countries.

The ultimate responsibility for fire safety within the home rests with the occupant the world over. This point is emphasized in a New Zealand fire service publication:

The recent series of tragic house fires has emphasized that it is the community and not the Fire Service which principally "owns" the fire problem. There are several elements of fire safety that only the building occupant can fulfill. Early detection, safe exit when fire occurs, effective fire prevention measures, safe suppression of fires still in the earliest stage of development, rapid and clear communication of the alarm of fire through the 111 telephone system, are all examples.[21]

Another New Zealand report indicates that the advent of furniture involving plastics and polycarbons has decreased the likely flashover time in a residential room from between 12 to 15 minutes with furniture constructed of wood and other natural

Fires in residential buildings cause nearly 90% of all fire fatalities in New Zealand.

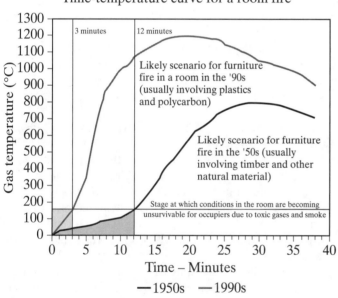

Flashover:

1950s: 12–15 minutes
1990s: 3–7 minutes
Less time to escape
No time for rescue

FIGURE 12.2 ◆ Research in New Zealand indicates a major change in flashover time as a result of changes in furniture construction. (Way Forward New Zealand, Fire Commission)

materials as used in the 1950s to 3–7 minutes with today's furnishings. This provides less time to escape and no time for rescue. Ceiling level temperatures are also increased materially by this change.[22]

Intensified fire prevention activities in the United States received international recognition in a resolution proposed for the European Parliament. One of the provisos stated that "the rewards have been high for Governments, such as the United States of America, which have organized a determined campaign to cut fire costs".[23]

◆ CANADIAN FIRE SAFETY PROCEDURES

The Canadian Department of National Defence, through its Fire Marshals' organization, regulates fire safety for military property and operations. Fire safety in other federal governmental buildings is under the control of the Occupational Safety, Health and Fire Prevention group of the Department of Labor. Retroactive sprinkler installations have been provided in governmental high rise buildings.

The Learn Not To Burn program of the National Fire Protection Association was introduced in Canada on a nationwide basis in 1993 under the auspices of Fire Prevention Canada. Canada is believed to be the only country in the world with adoption of this fire safety program in all schools.

Fire Prevention Canada is an organization that was formed in 1976 by the Council of Canadian Fire Marshals and Fire Commissioners and the Canadian Association of Fire Chiefs. It is the primary organization for the promotion of fire safety on the national level. The Canadian Association of Fire Chiefs plays a major role in fire service activities in that country.

Each of the ten provinces and two territories in Canada has a fire marshal or commissioner who heads up fire safety programs at that level of government. A number of these offices have a considerable amount of authority; in fact, the Ontario Fire Marshal is responsible for directing local governments to enhance fire protection capabilities when found to be inadequate. The Council of Canadian Fire Marshals and Fire Commissioners, which is comprised of the individuals previously noted plus the Department of Defence Fire Marshal, also publish the annual report on national fire statistics. As with other fire safety documents in Canada, both French and English language versions are available.

Smoke alarm use parallels that of the United States. These devices, plus extensive fire safety education and other measures, have cut average annual fire deaths in half, from 800 to 400, in recent years. This is especially noteworthy when unchanging severe winter climatic conditions are considered. Canada is experiencing the same problems with smoke alarm maintenance as in the United States.

The National Building Code of Canada and the National Fire Code of Canada serve as the foundation for fire safety engineering and fire safety enforcement throughout the country. Both are model codes developed by public interest committes of the National Research Council, a crown corporation responsible to the federal Department of Industry. These codes are adopted at the provincial or territorial level, usually without modification. Research activities of the National Research Council in fire safety are discussed in Chapter 13, as there is extensive interaction between the United States and Canada in this function.

Fire services at the local level operate in a fashion similar to those in the United States, with career, volunteer, and combination-type departments being utilized. Fire prevention activities are also carried out at this level. In recent years, there have been some major consolidations of local government in Canada with concurrent fire service amalgamation.

Canadian firefighters are also involved in fire prevention activities. As an example, departments in the greater Montreal area are using this technique. Montreal's program is described as follows:

> In addition, firefighters continue to carry out their prevention visits to various districts of the city. These visits provide the opportunity to exchange information with the public and detect any situations which may contribute to increasing the risk of fire.

> These various types of public contact enable firefighters to exercise their leadership among the public in the field of promoting self-protection against fire (smoke detectors, fire extinguishers, etc.).

> In this regard, since 1987 each shift of firefighters is required to contribute twenty-four and one-half hours of fire prevention work per week, over and above regular public fire prevention campaigns conducted by about forty inspectors, many of whom are women.[24]

VARIATIONS IN FIRE CODE APPLICATIONS

A major difference in philosophy exists between the United States and other countries. The United States is almost unique in having a system in which codes vary from

FIGURE **12.3** ◆ In 1929 70 children died in a panic situation in the Glen Cinema in Paisley, Scotland (on right). This event is described in Chapter 8. *(Paisley Local History Museum)*

one jurisdiction to another. In most other countries, fire safety standards, whether strong, moderate, or weak, are uniform within a given country. There may be minor differences from city to city; however, the general application of the code is usually the same throughout a given country. Application is much more uniform in Canada than in the United States.

In the United States, a city may have a very strong fire prevention building code, yet the unincorporated areas of the same county may have no code. There are many examples of hazardous operations that would not be permitted within the city limits but are carried out with no problem in adjacent unincorporated areas.

Building construction, sales of hazardous materials, and dispensing operations vary considerably across municipal boundaries throughout the United States. A visitor from another country has a difficult time understanding why a hazard is treated differently on one side of a municipal boundary than on the other because the immediate surroundings, people, activities, etc., appear to be the same.

VARIANCES IN PRACTICES AND PROCEDURES

In the United States there are examples of variances in practices and procedures among states. These differences appear not only in fire prevention and building codes, but in other fire-related measures as well. For example, one municipality may have a program for fire safety training for school children that is not conducted in neighboring

Figure 12.4 ◆ Fire prevention in international port cities such as Duluth, Minnesota, imposes special concerns such as unique cargoes and fire protection equipment.

communities. It is a matter of chance as to whether or not the school-age child has the benefit of fire prevention education in the classroom. The National Commission on Fire Prevention and Control looked on this variety in fire prevention efforts as a major problem.[25]

Fire prevention practices and enforcement procedures generally reflect the history of the nation as far as governmental control is concerned. Fire prevention practices in the United States probably are not uniform partially as a result of the experiences and attitudes of those who founded our government. Many immigrants came to North America in an effort to get away from extreme governmental regulation. As a result, the only requirements imposed by the original government in this country were those considered absolutely necessary for survival.

As urban areas developed, fire prevention and building code requirements became necessary to ensure domestic tranquility and safety. Individuals living in less densely populated communities saw no necessity for regulations to control conditions they did not view as hazardous. Therefore, we find that a great deal of the fire prevention and building code coverage in the United States came about as a result of specific problems arising at given times. Many states still lack fire prevention and building codes covering all conditions that might exist within the state. An example is the control of liquefied petroleum gas by a state that does not deem it necessary to maintain regulations governing overcrowding of places of assembly.

Philip Schaenman's report, *International Concepts in Fire Protection: Ideas from Europe that Could Improve U.S. Fire Safety,* suggests that some European fire safety procedures are worthy of emulation in the United States and Canada.[26] The author visited a number of European countries in an effort to learn why fire statistics are so much more favorable on the other side of the Atlantic Ocean.

The report suggests that European citizens are more aware of the hazards of fire than are North Americans because of the major fires that have stricken their cities during wars and during the Middle Ages. They support stronger fire regulations and receive more instruction on fire safety from their parents. The report also suggests that these countries have stronger building and fire safety codes and that these codes are in effect in rural as well as urban areas; many rural areas in the United States lack such codes. (Insurance also plays a more prominent role than in the United States. Insurance carriers have plan-review authority and may, in fact, operate as government agencies.)

The extensive use of chimney sweeps in Europe may also help prevent fires in residences. In some countries, visits by chimney sweeps are mandatory. Consumer products are regulated more strictly, and electrical and gas-powered products cannot be sold unless tested and approved by the government.

Schaenman found that European fire officers generally have more education in fire protection and building technology than their counterparts in the United States. This means that they are more oriented toward plan review and code enforcement. He also found that firefighter injuries are fewer in Europe. Self-contained breathing apparatus has been widely used in Europe since well before World War II. Firefighters have ridden inside vehicles rather than on the tailboard for an equally long period of time.

Most European firefighters are trained in fire prevention and code enforcement in their recruit training programs, but such training is less common in the United States. Senior fire officers in Europe devote a much greater percentage of their time to fire prevention and code enforcement duties.

The report also notes that the Fire Research Station at Borehamwood, United Kingdom, has been switching its research emphasis from structures toward contents, reflecting a growing need to address fire problems related to consumer products rather than to construction features.

Schaenman has prepared a third report, *International Concepts in Fire Protection: New Ideas from Europe.* Fire safety efforts of the United Kingdom, The Netherlands, Austria, and Hungary are examined in this publication. Austria and The Netherlands are recognized as having among the lowest fire death rates in the world. Both have excellent fire safety education programs.[27]

A fourth report by Schaenman, *International Concepts in Fire Protection: Latin and Asian Nations,* describes fire safety practices in those areas of the world. This information is especially helpful to fire safety practitioners in communities having immigrants from these nations.[28]

◆ SUMMARY

In making comparisons of fire prevention programs in effect on this continent with those in other parts of the world, there are probably as many differences as there are similarities. Standards of living, customs, dollar valuations, use of electricity, even smoking habits vary considerably from country to country. An intangible element that makes a great difference is the general adherence to constituted authority in many other countries. Travel restrictions, different kinds of heating devices, and methods of building construction are further examples of factors that must be taken into account in analyzing fire prevention practices and comparing results.

Some of the variables that affect statistics are diversified emergency services in the various countries; differences between the North American and European fire service systems with respect to placement and number of fire stations and amount of fire control equipment at each; and the great use of noncombustible outer wall construction in European countries.

Psychological factors also have a bearing on the fire situation in a country, and more research in this area is needed. It is not uncommon, for example, for Europeans to live out their entire lives in the same houses in which they were born. These individuals would be more apt to value the home and its contents and take greater precautions against the threat of fire than someone who moves about from dwelling to dwelling. The weight of authority in inspections also undoubtedly makes a difference in the occurrence of fires in a given country.

A major difference in philosophy exists between the United States and other countries with regard to the application of fire codes. In most other countries, fire prevention and building codes are as a rule uniform within the country and are usually enforced by a national agency. In the United States, many variations in code applications exist, in neighboring municipalities as well as from state to state, and enforcement is often left to the jurisdiction concerned. Canada's procedures are much more uniform than those of the United States.

Comparisons of our fire prevention practices with those of other countries will provide us with some valuable information. We can, in fact, learn a great deal from such studies and may find that it is highly desirable to include some of the better features of foreign programs in our own.

■■■

Review Questions

1. Which is not a factor that affects the fire situation in a given country?
   a. standards of living
   b. customs
   c. fire department structure
   d. use of electricity
   e. smoking habits
2. What country has a somewhat similar fire loss to that of the United States?
   a. United Kingdom
   b. Canada
   c. The Netherlands
   d. Czech Republic
   e. Austria
3. What country has the most building fires per 1,000 people in the modernized world?
   a. United States
   b. United Kingdom
   c. Canada
   d. The Netherlands
   e. France
4. What country has nearly five times the number of fire fatalities as the United States?
   a. Canada
   b. United Kingdom
   c. France
   d. Spain
   e. Russia
5. Monetary loss per fire is generally greater in other countries than the United States and Canada because of all of the following except
   a. proper phones in the United States and Canada
   b. fewer extinguishing systems in the other countries
   c. stations are closer together in the United States and Canada

   d. North Americans are less reluctant to call the fire department
   e. fires burn hotter in other countries
6. Which country or countries have recently experienced an upward trend in arson?
   a. United States
   b. United Kingdom
   c. Japan
   d. b and c
   e. a, b, and c
7. The single-engine substation is less common in
   a. the USA
   b. Canada
   c. Europe
   d. none of the above
8. Approximately _____ of the 18,000 members of the Tokyo Fire Department are devoted full time to fire prevention duties.
   a. 2 percent
   b. 5 percent
   c. 8 percent
   d. 10 percent
   e. 15 percent
9. The fire service in the United Kingdom is primarily organized on a _____ basis.
   a. county-wide
   b. city
   c. township
   d. state-wide
10. Fire Prevention Canada is an organization formed in
    a. 1968
    b. 1976
    c. 1981
    d. 1985
    e. 1989

■■■

Answers

1. c
2. b
3. a
4. e
5. e

6. d
7. c
8. d
9. a
10. b

■ ■

Notes

1. The Geneva Association, *World Fire Statistics No. 19,* Table 4 (October 2003).
2. Ibid.
3. Ibid.
4. "Russia Dorm Fire Kills 36" *Gainesville (Fla.) Sun* (November 25, 2003).
5. Personal communication to the author from the National Fire Protection Association, 1994.
6. Philip S. Schaenman, *International Concepts in Fire Protection: New Ideas from Europe,* p. 9.
7. Percy Bugbee, *Men against Fire: The Story of the National Fire Protection Association,* 1896–1971 (Boston: National Fire Protection Association, 1971), p. 15.
8. Ibid., p. 19.
9. Henry G. Hodges, *City Management Theory and Practice of Municipal Administration* (New York: F. S. Crofts and Co., 1939), pp. 486–488.
10. Boris Laiming, "Twentieth Century Fire Protection: History, Highlights, and Trends," mimeographed paper, Table 4, 1954.
11. D. I. Lawson, "Fire Protection in the USSR," *FPA Journal,* No. 64, Supplement 1 (London: Fire Protection Association), pp. 8–9.
12. Boris Laiming, "Twentieth Century Fire Protection, History, Highlights, and Trends," Table 5, 1954.
13. NFPA Fire Analysis and Research, *Intentional Fires and Arson,* Executive Summary (Quincy, Mass., 2003).
14. Percy Bugbee, *Men against Fire,* pp. 35–36.
15. Jerry G. Knight, "From St. Petersburg to St. Petersburg," *Fire Chief* (August 1993).
16. *White Book on Fire Service in Japan* (Tokyo: Japan Fire Protection Association, 1982).
17. Philip Schaenman and Edward Seits, *International Concepts in Fire Protection: Practices from Japan, Hong Kong, Australia and New Zealand* (Arlington, Va.: TriData Corp., 1985).
18. Jerry Senk, "The London Fire Brigade," *The Voice* (July/August, 1992).
19. Personal communication to the author from Dennis Ricketts, HM Fire Service Inspectorate, London, United Kingdom.
20. Ibid.
21. *The Way Forward: Modernization of New Zealand's Fire Service* (Petone, New Zealand: New Zealand Fire Service Commission, 1998).
22. *Way Forward* (Petone, New Zealand: New Zealand Fire Service Commission, 1997).
23. Resolution submitted by Sir Christopher Prout to European Parliament 1994 as reported by World Fire Statistics Centre, London.
24. *Changing for the Future* (Montreal, Quebec: Service de la prévention des incendies de Montreal, 1989), p. 12.
25. U.S. National Commission on Fire Prevention and Control, *America Burning* (Washington, D.C.: U.S. Government Printing Office, 1973).
26. Philip S. Schaenman, *International Concepts in Fire Protection: Ideas from Europe that Could Improve U.S. Fire Safety* (Arlington, Va.: TriData Corp., 1982).
27. Philip S. Schaenman, *International Concepts in Fire Protection: New Ideas from Europe.*
28. Philip S. Schaenman, *International Concepts in Fire Protection: Latin and Asian Nations* (Arlington, Va.: TriData Corp., 2000).

Fire Prevention Research

13 CHAPTER

Until the 1970s there had been a limited amount of research in the field of structural fire safety and prevention. On the other hand, there had been some excellent research efforts in forestry fire prevention.

The term *research* is defined as "careful, systematic, patient study and investigation in some field of knowledge, undertaken to discover or establish facts or principles."[1] A great deal of research effort was expended in both structural and forest fire prevention during the last quarter of the 20th century.

The need for structural fire prevention research is exemplified by the experience of a Coast Guard petty officer who was assigned to waterfront fire prevention responsibilities in Seattle during World War II. During one week the assignment was to post fire prevention posters in all waterfront facilities in Seattle. The following week the assignment was to conduct fire prevention inspections of the same facilities. The question later arose as to which week was the most effective as far as utilization of time for fire prevention was concerned. Were they of equal value, or was one more productive than the other? Unfortunately, little research has been done to provide answers to such basic questions.

A revealing research project was conducted in the 1950s by the Ithaca, New York, fire department. The department was concerned with the great number of fires occurring in the city as a result of the use of portable kerosene heaters and decided to have feature articles written for the local newspaper to depict dangers inherent in the use of these devices. The articles were amply illustrated and were designed to discourage the use of this equipment except under safe conditions. They learned that the newspaper campaign was of practically no value in reducing fire losses.

As a result of this experience, a survey was conducted to determine the newspaper reading habits of the individuals with the portable heater problem. An analysis of the city's fire records showed that practically all of the fires involving portable heaters occurred in lower-income residences. Further research indicated that individuals residing in these residences had reading habits that were quite different from those of individuals with higher income levels. Most of the former group did not read the feature page of the newspaper but were concerned primarily with the help wanted page, the comics page, and in some cases the sports page. This research was responsible for changing the direction of the fire prevention program so that the message was seen by the people who were most directly affected by this problem. This is an example of local research that can be most helpful to a fire department.

◆ U.S. FOREST SERVICE FIRE PREVENTION RESEARCH

The Forest Service of the U.S. Department of Agriculture has carried out a number of excellent fire prevention research projects. Some of these were conducted by Forest Service personnel; others were conducted by university personnel under contract to the Forest Service. These projects continue to guide Forest Service procedures, and additional research in these basic areas does not appear to be necessary.

More recent research has addressed prescribed burning, wildland urban interface situations, and related issues. Recent research is focused upon the concept of removing fuel rather than that of removal of fire causes. Basic cause removal efforts continue to be based on earlier research described next. Cause prevention research is emphasized in this chapter because of its relationship to structural fire prevention.

SMOKING REGULATIONS SIGNS

The Pacific Southwest Forest and Range Experiment Station of the Forest Service conducted a research project pertaining to no-smoking signs.[2] In many of the Southern California mountains, there is a prohibition against smoking while traveling in a vehicle through the area. These restrictions apply to some high-speed multilane highways. Signs are provided at the entrance to these restricted zones; in some cases, areas

FIGURE 13.1 ◆ This fire watch tower behind Charleston, South Carolina's historic fire headquarters once provided service to the city.

are provided in which a motorist may stop and smoke. A sign that showed a smoldering cigarette with the word NO was compared to a sign with a picture of a smoldering cigarette with no wording but with a slash across the cigarette. The signs were very effective in attracting the eye of motorists. Eighty-five percent of those passing the one with the word and the picture recalled the message. Seventy-three percent of those who saw the sign with no wording recalled the message. This indicates the desirability of illustrated signs with a very limited amount of wording as a public warning.

A similar survey was done with signs indicating areas where motorists might stop and smoke. Again it was found that a sign with OK on it was superior to a sign that merely showed a lighted cigarette with no wording whatsoever and without the slash mark. An interesting feature of the latter survey was the tendency of people stopping at the smoking area to steal the signs. A great number of the signs were apparently removed by youthful sign collectors.

The Pacific Southwest Station in Berkeley also conducted a survey regarding knowledge of smoking regulations in the national forests of Southern California.[3] Researchers were interested in determining the accuracy of people's understanding of forestry regulations. They found that individuals living close to the forests were generally more familiar with the regulations than were those who lived at distant locations. They also found a direct relationship between frequency of visits to the forest and possession of correct information. Those who visited the forest ten or more times understood the regulations almost twice as well as those who were traveling in the forest for the first time. The researchers found that approximately one of every four smokers traveling in the forest does not understand prevailing smoking regulations. This clearly indicates the need for disseminating specific information.

CHILDREN AND MATCHES

The U.S. Forest Service, in cooperation with the University of Southern California and the California Division of Forestry—now the Department of Forestry and Fire Protection—also conducted research into "children with matches" fires.[4] This research was conducted to determine problems with and possible solutions for these fires. Approximately three-fourths of the fires attributable to juveniles were started by children 10 years of age or younger. Twelve percent of the 484 fires studied were started by children under 5 years of age; 32 percent were started by children between 5 and 7. Children 8 to 10 caused 28 percent of the fires studied, and the remaining fires were caused by children 11 or older. Ninety-two percent of the offenders were boys.

The direct result of this research was a realization that fire prevention education should include preschool and early elementary grades if there is to be abatement of the fire setting problem. Very young children might be most effectively reached through popular television programs for their age level. The research bears out the need for interchange of information among investigative agencies handling the fire problem within a given area.

PUBLIC AWARENESS OF FIRE PREVENTION IMPORTANCE

An earlier study by the Pacific Southwestern Forest and Range Experiment Station addressed the fire prevention knowledge and attitudes of the residents of Butte County, California.[5] This county was selected because of the high percentage of the population that goes into the wildlands of the county at some time during the year. All social and economic factors regarding the surveyed residents of Butte County were analyzed. These included age, marital status, education, job status, type of work, sex, race, current and previous places of residency, or migratory status. The survey indicated that fire prevention knowledge was higher among individuals who used the wildland for either pleasure or work on a frequent basis. It also indicated a definite relationship between socioeconomic variables and experience with forest fires. It noted that frequent users scored best in both knowledge and attitude scales. There was a strong and positive feeling toward fire prevention on the part of most respondents. This feeling was intensified by personal experience with wildfires and the extent of experience in wildland areas.

There were also noticeable groupings of individuals who were frequent users of the wildlands who had little knowledge of or interest in fire prevention problems. These included high school dropouts. Contrary to general opinion, females were equally represented in this group.

RESEARCH IN FIRE PREVENTION PROMOTION AND ENFORCEMENT

The Forest Service also conducted a survey in Southern California to determine the best means of promoting fire prevention.[6] Researchers found that a successful fire prevention campaign contains two elements: It should teach *why* individuals must be careful with fire, and it should also teach *how* we should be careful with fire. An attempt was made to determine what the public already knew about fire prevention. Researchers went into the forests to study fire prevention measures. They interviewed individuals who had been in the forests and questioned them about their fire prevention practices.

They found that men generally scored better than women in fire prevention knowledge. They also found that people in the 25 to 50 age group had the most correct information on fire prevention and that those in the under 25 age group generally lacked information. In the over 50 age group, the problem was with misinformation.

FIGURE 13.2 ◆ Haz-Mat duties of the fire department can be explained to citizens at public events. *(Capt. John Turner, Louisville, Kentucky, Fire Dept.)*

The study further revealed that individual contact taught the most, followed by information contained in written material. The third most important means of conveying fire prevention information was through radio and television, which were useful in conditioning attitudes. Individual contact was found to be useful in pointing out specific fire prevention procedures. Written material was also found to be helpful in the latter category.

ENFORCEMENT OF FIRE LAWS

The same study by the Forest Service also addressed the problems of fire law enforcement. The researchers interviewed individuals who had been charged with violations of fire prevention regulations. They also interviewed individuals who had followed favorable practices in the use of campfires in forested lands. This study was quite comprehensive and included a considerable amount of review of case studies and interviews with appropriate individuals. The results of this survey were revealing and gave indications of problem areas that relate to structural as well as forest fire prevention.

The survey revealed that the average violator in the forested area was an honest, intelligent individual who was aware of the importance of fire prevention but who had not read or understood the specific regulations. The survey further indicated that the majority of people believed that fire prevention regulations are not generally known and recognized and that fire prevention laws and regulations are as important as traffic regulations. Most of those interviewed stated that they would observe both types of regulations equally. Most of the nonviolators interviewed said that they saw the need for more enforcement of forest fire prevention regulations; less than one-half of the violators interviewed believed that less enforcement should be practiced.

A survey of the practices of the violators revealed that most did not realize they were doing anything incorrectly. They were following practices they thought were safe but that turned out to be violations of fire prevention regulations.

Both violators and nonviolators were asked the question "Have you received any traffic citations in the last two years?" In answering this question in the affirmative, violators outnumbered nonviolators three to one. This indicates the possibility of a close relationship between individuals who violated fire prevention regulations and those who violated traffic regulations and laws. It may further indicate that certain members of the public are indifferent to regulations that affect them directly and thus choose to ignore these requirements.

Another phase of the study examined enforcement efforts by individuals assigned such responsibilities. The assignment of personnel to fire prevention enforcement duties does not necessarily mean that prompt enforcement will be effected. The survey noted the need for full review of enforcement policies relating to fire prevention requirements and a review of the uses of the laws and regulations in effect. Certainly, there is little value to laws that are on the books but that are not enforced.

MEDIA USES IN FIRE PREVENTION PROMOTION

The study included a detailed evaluation of the "senders," who might be defined as the sources of information in the fire prevention field, as opposed to the "receivers." This included a detailed survey of all media used for disseminating fire prevention information.

The study surveyed the degree of credibility placed on the senders by the receivers. Some individuals are not inclined to believe information on television, on radio, or in a newspaper. Such prejudices were evaluated so that people responsible for publicizing fire prevention might be guided in using the proper media. The program also included a survey of media employed for fire prevention efforts with relative values assigned. An evaluation was made of the credibility of media, generally without specific reference to credibility of the particular receivers who were interviewed. This involved a general survey of public information to determine appropriate use of the different methods. The methods were studied from the standpoint of applicability to the public. This portion of the survey was primarily related to forest users who were considered as "the public" for the purpose of the survey.

The messages given out in fire prevention education efforts were also studied in detail for composition and general attractiveness. Message content was analyzed, with emphasis on the semantics of the message. The word *watershed,* for example, was found to connote to many members of the public a structure in which water is stored. This indicated the need for the use of a word having more clear meaning to the majority of people.

The survey checked the use of the words *closed area.* It found that this term also had little meaning to the average person who was not fully familiar with forestry procedures. A number of people were not sure of the meaning of the term *limited access* and could be easily misled by a sign bearing it. The public had varying opinions of the meanings of *highway* and *parkway.*

PUBLIC FIRE EDUCATION RESEARCH

Several excellent fire prevention research projects were carried out with the support of the Public Education Office of the National Fire Prevention and Control Administration, now the U.S. Fire Administration. These projects were designed to shed light on problems in public fire education and to suggest possible directions for improvement.

A preliminary report was submitted by Whitewood Stamps, Inc., as a fire education research study.[7] The study was designed to place *fire education,* a term encompassing both fire prevention and fire reaction, in context and to delve into high risk as it relates to people, environments, and situations. It also examined the processes by which fire education has been developed.

The report brought out the fact that fire education is a part of the public health and education fields as well as the fire protection discipline. It noted that some excellent fire education programs are carried out by public health and education personnel.

Researchers recognized that "the success of fire education messages depends on the extent to which local conditions are taken into account—e.g., cultural influences, values and beliefs; and types of hazards encountered."[8] They also recognized that the "success of fire education can be limited by the extent of people's control over their economic resources, physical capacities, and the inherent safety of consumer products, as well as by selective perception and competing priorities."[9]

High-risk groups, which often include the young, fire setters, adults, and the older adults, were discussed in relation to ignition problems and to death and injury potential. The term *fire setters* normally refers to juveniles rather than individuals normally thought of as arsonists.

Researchers found that scalds rather than open flames are the greatest burn-related threat to children under three. Supervision and a safe environment were found to be equally necessary to prevent fire-related accidents to children.

Adults were found to be in a high-risk category in relation to ignitions but are in a low-risk category for death and injury because of their competency in handling fire situations.

To the contrary, the elderly tend to be a low-risk group in relation to ignitions but are in the high-risk category for death and injury. A distinction must be made, the researchers believe, between self-sufficient older people and those needing home care and supervision.

Current public education programs were examined and categorized into three types: community action programs, in-school programs, and mass media programs. Five elements appear to be essential to the success of fire education programs: the identification of local fire problems, selection of education strategy, design of materials, implementation, and evaluation of the education program.

Effective use of mass media, it was found, requires knowledge of the value structure of the target population. If messages are to be understood correctly, they must be framed in terms of existing beliefs and attitudes. Localism should be emphasized in the design of educational materials. The importance of having an evaluation procedure was stressed.

This project was followed by the development of a manual entitled *Public Fire Education Planning,* also prepared by Whitewood Stamps, Inc., under the auspices of the then National Fire Prevention and Control Administration.[10] This manual delineates the steps that should be followed in implementing a public fire education program in a community, county, or state.

The manual suggested as a starting point the establishment of responsibility for fire education planning. This includes the assignment of administrative, policy, and staff responsibilities for decisions and activities.

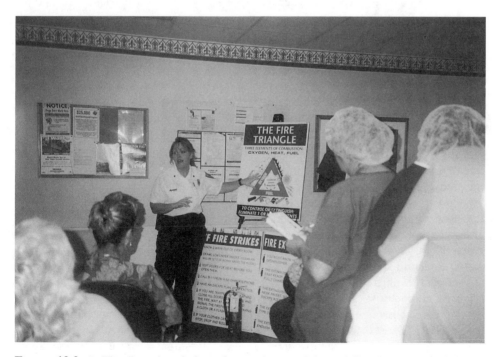

FIGURE 13.3 ◆ The fire triangle is an integral part of the public educator's training tools. (*Lakeland, Florida, Fire Department*)

New Orleans was the site of two meaningful research projects in fire prevention carried out with federal support. The first study was conducted by Planning and Research Associates, Inc., of Baton Rouge, with the assistance of the New Orleans Fire Department. It was entitled *The Human Factor in High Fire Risk Urban Residential Areas.*[11] The second study was entitled *Fire Prevention Program Planning in a Social Action Perspective: An Exploratory Study of Knowledge, Attitudes, and Leaders in a High Fire Risk Area of New Orleans, Louisiana.*[12] This survey was conducted by the same team.

Both studies took place in a high fire risk urban residential area in the central part of New Orleans. The area had around 25,000 inhabitants, of whom 91 percent are black. The predominant housing unit is the duplex, although there are some single-family and high- and low-rise housing projects in the area. Ten percent of the housing units were occupied by their owners at the time of the study.

The goal of the first study was to determine the economic, cultural, and demographic variables related to fire occurrence in a high fire risk urban residential area. The study resulted in the following summary of findings:

1. Preliminary investigation determined that the types and kinds of fire in the study area were similar to those found in high-risk areas in other cities.
2. Interviews with firefighters and residents indicated that fires were primarily caused by careless behavior, such as smoking in bed, placing flammables near open flames, and leaving food unattended on stoves; and by tolerant attitudes toward trash and debris and behavior of juveniles and vagrants.
3. Analysis of the data obtained suggested that the human factor, expressed in terms of a subculture, would have to be considered in planning programs of fire prevention. Fire proneness is the result of these specific subcultural characteristics:
   - lack of a strong spirit of community,
   - lack of fire safety training,
   - lack of practice of fire precautions,
   - lack of fire fighting facilities, and
   - an attitude of fatalism toward fire.[13]

The study found that many children in the area used fire as a plaything and that this practice was not strongly discouraged by parents and other adults. It also noted that the area under study tended to attract all types of vagrants, many of whom frequented vacant buildings. These individuals seemed to be more readily accepted in the study neighborhood than they would be in many other areas.

Preventive programs for high fire risk areas will need to be especially designed, the survey indicates. There is a need to change ways of thinking about fires in such areas; there is also a need to identify and utilize proper communication channels to reach the target population, according to the survey report.

The second study attempted to determine what social action could be initiated at the local level to reduce the rate of fires and fire losses in residential areas with high-risk characteristics.[14]

Research personnel found that one-third of the residents interviewed in the target area had experienced a fire in their dwelling at one time or another. Of these, half had experienced a fire within the previous ten years. Residents experiencing fires seem to have "learned a lesson" from their particular fire but not a general lesson in fire prevention. They did not gain a full appreciation for the broad concepts of fire prevention but only gained an appreciation of preventive measures relating to their personal fire experience, whether it be careless use of smoking material or leaving food on the stove unattended.

Personnel assigned to the survey found that there is often a distrust of outsiders in neighborhoods of this type. There is also, they found, a fear of involvement and distrust of "programs" as well as a casual attitude toward fire prevention. They concluded that the conditions found in New Orleans were generally similar to those of high fire risk areas in other cities.

Specific conclusions of the second survey include a finding that the residents of the area studied have a strong awareness of fire as a problem. They also found a strong belief among the residents that the fire department should take the responsibility for fire prevention programs at the city level. They believed that neighborhood programs in fire prevention could be performed by other public agencies as well, such as schools and social service agencies.

The survey found that the mass media, especially television and radio, have a potential for reaching residents with a fire prevention message. The leadership structure in the study area is complex, and this force is not as apparent as in more cohesive neighborhoods. Local leaders, it was found, are primarily church leaders, government employees, and elected officials.

Surveyors contacted the area leaders and found that these individuals were well aware of the fire problem and the need for fire prevention programs. These individuals indicated an apparent willingness to participate in such activities.

The New Orleans projects gave fire prevention personnel some excellent information on which to base programs for high fire risk areas. The results can be of value in designing programs for such areas.

◆ NATIONAL INSTITUTE OF STANDARDS AND TECHNOLOGY RESEARCH

Research activities of the Building and Fire Research Laboratory (BFRL) of the National Institute of Standards and Technology (NIST) in Gaithersburg, Maryland, include building materials; computer-integrated construction practices; fire science and fire safety engineering; and structural, mechanical, and environmental engineering. Products of the Laboratory's research include measurements and test methods, performance criteria, and technical data that support innovations by industry and are incorporated into building and fire standards and codes. The Laboratory's technical program is carried out by three divisions: Materials and Construction Research, Building Environment, and Fire Research. The Fire Research Division develops, verifies, and utilizes measurements and predictive methods to quantify the behavior of fire and the means to reduce the impact of fire on people, property, and the environment. The Fire Research Division provides leadership for advancing the theory and practice of fire safety engineering, fire fighting, fire investigation, fire testing, fire data management, and intentional burning.

In order to reduce the risk of flashover, the Fire Research Division continues to conduct research to assess the most promising technologies for building less flammable mattresses. Fires in which bedclothes or a mattress are the first item ignited result in hundreds of deaths each year. NIST is characterizing the hazard represented by a given bed fire size in an attempt to assess the potential for relating fire size reduction to lives that may be saved. The Fire Research Division is also working with the

Consumer Products Safety Commission (CPSC) to define a test method by which CPSC will be able to assess the flammability of mattresses.

A key factor in reducing firefighter and building occupant fatalities and burn injuries is to develop advanced technologies for the fire service. NIST is developing standard measurement methods to advance a heat transfer model for firefighter protective clothing under wet and dry conditions. The model will be used to generate training tools to allow firefighters to visualize the limits of performance of their protective gear. Working with a consortium of fire alarm manufacturers, the Fire Research Division is developing a wireless means to deliver timely emergency information about conditions inside of buildings. Predictions of developing hazards to first responders before they arrive at the scene will allow firefighters to better understand the optimum plan for rescue and suppression operations. Interacting with the industry, BFRL will focus on expanding the capabilities of firefighter equipment, such as thermal imagers or personal alert safety system (PASS) devices, for improved sensing for fire hazard and fire rescue applications.

NIST is continuing to develop the Fire Dynamic Simulator, a computer model which incorporates state-of-the-art fluid dynamics, physics, and chemistry for fire reconstruction. Fire service training instructors are including simulations in training programs for firefighters. Fire protection engineers use the model to develop more fire safe structures. The Fire Dynamic Simulator has provided insight into a number of fire recreations including a townhouse fire (Cherry Road, Washington D.C.), a duplex residential structure fire (Keokuk, Ia.), a restaurant fire (Houston, Tex.), office buildings fire and collapse (New York, N.Y.), and a nightclub fire (West Warwick, R.I.).

Personnel of the Laboratory are active in many standards-making organizations. This is one of the methods for transferring technology developed by the Laboratory into useful applications. A summary of the goals of the Laboratory is contained in Chapter 10.[15]

◆ UNIVERSITY RESEARCH

Another example of research is the development of the Kidsmart vocal smoke alarm, designed to waken children from deep sleep with a verbal command in the parent's voice coupled with a tone siren. This device was designed by three M.B.A. students at the University of Georgia and won the grand prize at the 2003 annual competition of the intercollegiate business plan corporation at the University of Texas at Austin.

Pilot testing of the Kidsmart vocal smoke alarm shows that it is effective in awakening children. Their response usually occurs within 15 seconds of the commencing of vocal instructions by the parent.[16]

◆ CANADIAN RESEARCH ACTIVITIES

The Institute for Research in Construction of the National Research Council of Canada was established to serve the needs of Canada's construction industry. This industry is Canada's largest, representing almost 20 percent of the gross national product.[17] Many of those products are exported to the United States.

The National Fire Laboratory of the Institute is responsible for conducting research aimed at reducing life and property losses by fire. The Laboratory was established as a direct result of a major ship fire in the port of Toronto. Priority is given to subjects related to the National Building and Fire Codes of Canada and to assistance to the building industry through research into fire reaction of building materials and assemblies. The Laboratory has a close working relationship with the Building and Fire Research Laboratory in the United States.

The Canadian fire research facility located near Ottawa is probably the finest such facility in the world. It includes a 10-story tower in which high-rise fire safety studies can be conducted. It has been used to conduct evacuation research for such buildings.

Research activities of the Laboratory include fire risk assessment, costs of fire safety including public protection, effectiveness of structural fire protection including fire barriers, and high efficiency water-based fire suppression systems. Other aspects of fire protection are also studied.

Straw-built construction is on the rise in the United States and Canada. It has a very high energy efficiency rating resulting in a savings in heating costs. The National Research Council of Canada, as an example of its work, tested tightly compacted bales of straw. Researchers found that the fact the bales are closely compacted greatly reduces the airflow potential, thereby preventing fire buildup. Another advantage of straw-built construction is the absence of wood studs separated by open space, thereby not permitting upward fire spread. The compact straw bales preclude this possibility.[18]

◆ CODE ENFORCEMENT RESEARCH

The Urban Institute, the National Fire Protection Association, the National Science Foundation, and the U.S. Fire Administration combined forces to conduct extensive research into the area of fire code inspection practices.[19] Experts have long thought that fire code inspections result in fewer fires, lower fire losses, and fewer civilian casualties in properties covered by fire codes. They have also thought that fire codes prevent hazards and deficiencies that cause fires.

The study identified the fires and civilian fire casualties that may be prevented by inspection and measured the number of such fires and casualties. It also identified potentially important characteristics of fire code inspection practices and measured differences in practice from city to city. A representative group of 11 large cities was selected for in-depth study. Data from 6 other cities and 1 county were used for comparative purposes. Incendiary and suspicious fires were excluded because there is limited potential for preventing them through inspections.

Cities selected for the survey were Charlotte, North Carolina; Columbus, Ohio; Denver, Colorado; Houston, Texas; Indianapolis, Indiana; Jacksonville, Florida; New Orleans, Louisiana; San Jose, California; Seattle, Washington; Tucson, Arizona; and Wichita, Kansas. The fire code inspection practices analyzed were divided into four groups: measures of inspection frequencies; inspector preparation (training and experience); motivational practices, including techniques used to ensure compliance; and practices, including plan reviews, aimed at controlling building construction features related to fire safety.

The comprehensive report of this project contains several conclusions. Because these conclusions contain some excellent suggestions for fire safety inspection program management, they are included in summary form:

1. Cities that annually inspect all or nearly all inspectable properties appear to have substantially lower fire rates than do other cities. Cities in which a substantial share of the

inspectable properties go several years between inspections, or are not regularly inspected at all, tend to have higher fire rates.

2. Cities that use fire suppression companies for a large share of their regular fire code inspections appear to have substantially lower fire rates than do cities that use full-time fire prevention bureau inspectors exclusively. The probable reason is that cities using full-time inspectors exclusively often do not have sufficient personnel to make annual inspections of all inspectable properties, whereas cities that use fire suppression companies usually have the needed personnel.

3. Cities that assign inspection responsibility by geographic districts appear to be more successful in covering all inspectable properties. In the district-based approach, city blocks or streets are covered each year, and therefore each inspection cycle provides a new opportunity to locate an inspectable property that was not spotted before, either because it was not there in the previous cycle or because it simply was not noticed before.

4. There was no evidence that differences in fire rates are sensitive to other differences in the inspection practices examined: i.e., techniques used to prepare inspectors, to ensure thorough inspections, to persuade owners and managers to comply, to enforce compliance when voluntary compliance is not possible, or to control building features. The word *technique* is important here because action is being assumed in all cases; the variable is merely in methods or "techniques" used to achieve the goal.

5. Despite the almost universal belief that selling fire prevention is the key to success in obtaining compliance, inspectors were given little or no training or policy guidance on persuasive techniques.

6. Although the evidence did not show that any particular persuasive techniques were associated with lower fire rates, there was some evidence that the motivational and persuasive impact of inspections is more important than the direct removal of hazards. In inspectable properties in all communities examined, fires resulting from careless or foolish actions or electrical or mechanical failures (which are sometimes because of poor maintenance) greatly outnumbered fires resulting from visible hazards that inspectors are likely to remove directly. Fires of the first kind (which tend to be preventable primarily through educational and motivational efforts) typically constituted 40 to 60 percent of all fires in a particular community; fires of the second kind typically constituted 4 to 8 percent of all fires. (The remaining fires were of incendiary, suspicious, natural, or undetermined cause.) The most important difference in motivational and educational effects thus appears to be whether inspectors get around each year; if they do, they will have more effect than inspectors who are rarely if ever seen.

7. Deficiencies in building features and the absence of automatic fire protection equipment were factors in most incidents involving ten or more civilian deaths in inspectable buildings in the United States. Fire departments might well consider giving priority attention to building features most often implicated in fire involving large numbers of casualties. The lack of large loss of life in school fires since the early 1960s may be the result of the nationwide crackdown on school fire safety following the Our Lady of Angels fire in Chicago.

◆ NATIONAL FIRE ACADEMY RESEARCH

Students attending the Executive Fire Officer Program at the National Fire Academy in Emmitsburg, Maryland, are required to prepare and submit applied research projects. Abstracts of the following two projects represent applied research that contributes to the advancement of fire safety.

Abstracts

In December 1997, the City of Waterbury appointed a new fire chief. After his appointment, a number of changes occurred that would define how the Waterbury Fire Department would provide services to the community. The Fire Chief established a goal to increase the role of the fire department's fire fighting forces in fire prevention activities throughout the community. The problem was that the Waterbury Fire Department was not effectively using fire department personnel in delivering public fire safety education. Fire companies had been a small part of the public fire safety education program, but soon there would be an increase in their role in delivering these programs to the public. The purpose of this applied research project was to develop a plan to effectively use suppression personnel in the Waterbury Fire Department to provide public fire safety education. This research project used the evaluative methodology to examine what steps would be needed to develop a plan to use suppression personnel by the Waterbury Fire Department to effectively deliver the Waterbury Fire Department's Public Fire Safety Education Program. The research questions to be answered were: (1.) What procedures and methods should a fire department use to implement a public fire safety education program? (2.) What training will be required for the members of the Waterbury Fire Department Bureau of Emergency Services so that they may effectively deliver a public fire safety education program? (3.) What obstacles will the Waterbury Fire Department need to overcome in developing suppression personnel so that they can provide effective public fire safety education? A survey of suppression personnel assessed how those members viewed their role in public fire safety education. This research found that members of the Waterbury Fire Department were willing to conduct public fire safety education programs, but they felt that they were not prepared to do so. The survey indicated that additional training for personnel and additional public education resources were needed to effectively deliver the programs. The literature review examined what steps the Waterbury Fire Department would need to take as well as what obstacles the Waterbury Fire Department would need to overcome to implement the needed changes in using the suppression personnel more effectively in delivering public fire safety education. Recommendations included a greater commitment from the department, especially administrative and chief officers. The department must provide training that is appropriate for new recruit firefighters, firefighters and additional training for all company officers. The additional resources to be used by fire companies for public education programs must be purchased and maintained. These resources should be adequate in number and appropriate for each specific program.[20]

The problem was that the Mount Prospect Fire Department had not evaluated the use of civilian Fire Prevention Bureau personnel since they completely replaced uniformed personnel in 1995. The purpose of this applied research project was to evaluate the effectiveness of using civilians in fire prevention and develop a draft proposal on how to most effectively staff the Mount Prospect Fire Prevention Bureau. Action and evaluative research methods were used to answer the following research questions: (1.) What fire service changes are taking place that impact Fire Prevention Bureau staffing? (2.) How do similar suburban fire departments in the Chicago area staff their Fire Prevention Bureaus? (3.) What are the fire prevention bureau staffing problems encountered by fire departments surveyed in the Chicago area? (4.) What are the advantages and disadvantages of using civilians to staff Fire Prevention Bureaus? The research was accomplished through a literature review, telephone interviews, and surveys of similar suburban fire departments in the Chicago area. Literature searches were initiated at the National Emergency Training Center's (NETC) Learning Resource Center and the public library system of the City of Naperville, Illinois. Extensive searches were also conducted online through Internet search engines to identify current documentation of the research topic. The results of the research indicated Fire Prevention Bureau personnel need to be more technically trained. The research revealed fire departments in Chicago area staff their Fire Prevention Bureaus with civilians, uniformed personnel, or a combination of both. The results of the interviews revealed using civilians in a large municipal Fire Prevention Bureau is cost effective with longer longevity. The research indicated the Mount Prospect Fire Department does not have a Fire Prevention Bureau staffing plan. The researcher recommended adoption of the draft proposal to staff the Mount Prospect Fire Prevention Bureau in Appendix A.[21]

◆ OTHER RESEARCH ACTIVITIES

The Report of the Federal Emergency Management Agency, U.S. Fire Administration, *America at Risk, Findings and Recommendations on the Role of the Fire Service in the Prevention and Control of Risks in America; America Burning Recommissioned,* contains comments that are recommendations relating to research in the fire field. Issued May 2, 2000, the report noted the valuable research work being done by several federal agencies, including the Consumer Product Safety Commission and the National Institute of Standards and Technology, but called for more focused and coordinated efforts.

The first paragraph of the fire research recommendation sums up the problem:

FEMA should take a leadership role in setting agenda for research into fire and other risks for which the fire and emergency services communities have responsibility. As a first step, a reasonable set of priorities should be established for fire issues. Research agendas should be set with significant user input and influence. In addition, partnerships among NIST and other governmental, university, international and private research organizations can be utilized to develop research agendas that include issues connected with building codes and standards.[22]

An example of the type of research project referred to in the preceding above paragraph is outlined in this notice sent to Texas fire chiefs by their association in 2003:

Identification of Residential Fire-Injury Behavioral Risk Factors
The Department of Fire Protection Engineering at the University of Maryland is conducting research as part of a consortium, with support from the Center for Disease Control/National Center for Injury Prevention and Control. The principal goal of this 3-year research project is to identify specific residential fire and fire injury–related behavioral factors that are

amenable to intervention and prevention strategies. This information will be used to assist public health agencies and policy-makers in developing more cost-effective residential fire injury prevention programs. Specifically, this study seeks to collect data on residential fire-related injuries and fatalities by interviewing residential fire victims or their surrogates in order to obtain detailed information concerning the sequence of events and human behaviors that resulted in the casualty.[23]

◆ **SUMMARY**

Research programs carried out by the U.S. Forest Service have been directed toward efforts to prevent forest fires. Experiments have been made with different kinds of roadside fire prevention signs to determine their relative effectiveness. A similar project was designed to test the value of different kinds of no-smoking signs and to determine public understanding of smoking regulations. Other research studies have been made of fires caused by "children with matches," public awareness of the importance of fire prevention, public image of Smokey Bear and its effect on fire prevention, fire hazard inspection procedures, methods of fire prevention promotion, and problems of fire law enforcement.

Among areas in which additional research is needed is the problem of fires caused by children. More information is needed about the sociological and psychological factors that cause some children to set fires, and research may well point to the need to redirect fire prevention education programs for young children.

Researchers are leading the way to some scientific answers to the question of how to prevent fires, and recent trends indicate that research will become increasingly important in the future directions of fire prevention.

■ ■

Review Questions

1. "Careful, systematic, patient study and investigation in some field of knowledge, undertaken to discover or establish facts or principles" is
   a. discovery
   b. science
   c. physics
   d. research

2. _____ of people shown a no-smoking sign with the word NO and a picture recalled the message.
   a. 55 percent
   b. 65 percent
   c. 75 percent
   d. 85 percent
   e. 95 percent

3. According to a United States Fire Service study, approximately _____ of the fires attributed to juveniles were started by children age 10 and under.
   a. 65 percent
   b. 70 percent
   c. 75 percent
   d. 80 percent
   e. 85 percent

4. Successful fire prevention campaigns, according to the Forest Service, contain which two elements?
   a. why be careful and how to be careful
   b. when to be careful and how to be careful
   c. why to be careful and when to be careful
   d. where to be careful and when to be careful

5. According to the Forest Service survey, who scores better in fire prevention knowledge?
   a. men
   b. women
   c. children
   d. seniors

6. The term *fire setters* refers to
   a. all fire setters
   b. arsonists
   c. seniors
   d. juveniles
   e. teens

7. Fire proneness is the result of which subculture characteristic?
   a. lack of strong spirit of community
   b. lack of fire safety training
   c. lack of fire fighting facilities
   d. lack of practice of fire precautions
   e. all of the above

8. The National Fire Laboratory in Canada was established as a result of a/an
   a. hotel fire
   b. ship fire
   c. house fire
   d. school fire
   e. industrial fire

Answers

1. d
2. d
3. c
4. a

5. a
6. d
7. e
8. b

Notes

1. *Merriam Webster Dictionary* (Springfield, Mass.: Merriam Webster, Inc., 1995), p. 445.
2. William S. Folkman, "Signing for the 'No Smoking' Ordinance in Southern California," U.S. Forest Service Research Note, PSW-71 (Berkeley, Calif.: Pacific Southwest Forest and Range Experiment Station, 1965).
3. William S. Folkman, "Motorists' Knowledge of the 'No Smoking' Ordinance in Southern California," U.S. Forest Service Research Note, PSW-72 (Berkeley, Calif.: Pacific Southwest Forest and Range Experiment Station, 1966).
4. William S. Folkman, "'Children-with-Matches' Fires in the Angeles National Forest Area," U.S. Forest Service Research Note, PSW-109 (Berkeley, Calif.: Pacific Southwest Forest and Range Experiment Station, 1966).
5. William S. Folkman, "Residents of Butte County, California: Their Knowledge and Attitudes Regarding Forest Fire Prevention," U.S. Forest Service Research Paper, PSW-25 (Berkeley, Calif.: Pacific Southwest Forest and Range Experiment Station, 1965).
6. Lloyd M. LaMois, *Research in Fire Prevention,* Second General Session, Third Annual Governor's Statewide Conference on Fire Prevention (Annapolis, Md., 1961), pp. 37–43.
7. Whitewood Stamps, Inc., *Fire Education: People, Programs, and Sources,* prepared for the National Fire Prevention and Control Administration (Newton, Mass., 1975).
8. Ibid., p. 3.
9. Ibid.
10. Federal Emergency Management Agency, *Introduction to Fire Safety Education,* Appendix 1, 1985.
11. *The Human Factor in High Fire Risk Urban Residential Areas: A Pilot Study in New Orleans, Louisiana* (Baton Rouge: Planning and Research Associates, Inc., prepared for the National Fire Prevention and Control Administration, 1976).

12. *Fire Prevention Program Planning in a Social Action Perspective: An Exploratory Study of Knowledge, Attitudes, and Leaders in a High Fire Risk Area of New Orleans, Louisiana* (Baton Rouge: Planning and Research Associates, Inc., prepared for the National Fire Prevention and Control Administration, 1977).

13. *The Human Factor in High Fire Risk Urban Residential Areas,* 1976, pp. 1–2.

14. *Fire Prevention Program Planning in a Social Action Perspective,* 1977, p. 20.

15. Personal communication to the author from representative of Director of Building and Fire Research Laboratory, National Institute of Standards and Technology, 2003.

16. *Texas Fire Chiefs Association Weekly Report* (May 24, 2003).

17. National Fire Laboratory of the National Research Council, Canada (Ottawa, 1993).

18. Jonah Marc O'Neil, "Fire Resistance of Straw Bale Houses," *Canadian Fire Chief* (Ottawa, Spring 2003), pp. 18–23.

19. *Fire Code Inspections and Fire Prevention: What Methods Lead to Success?* The Urban Institute and National Fire Protection Association supported by National Science Foundation and U.S. Fire Administration (Washington, D.C., 1978).

20. George Klauber, *Firefighters to Fire Preventers: A Change for the Waterbury Fire Department,* an Applied Research Project submitted to the National Fire Academy as part of the Executive Fire Officer Program (Emmitsburg, Md., 1999).

21. R. Paul Valentine, *Retention of Civilian Fire Prevention Personnel,* an Applied Research Project submitted to the National Fire Academy as part of the Executive Fire Officer Program (Emmitsburg, Md., 2000).

22. Federal Emergency Management Agency, U.S. Fire Administration, *America at Risk, Findings and Recommendations on the Role of the Fire Service in the Prevention and Control of Risks in America* (Washington, D.C., 2002), p. 22.

23. Identification of Residential Fire-Injury Behavioral Risk Factors notice by Texas State Fire Marshal's Office, May 2003.

CHAPTER 14

Proving Fire Prevention Works

Fire prevention works; however, this must be proven year after year. The preceding chapters have illustrated many examples of successful fire prevention programs.

In 1928, Chief W. D. Brosnan of Albany, Georgia, spoke on this subject at the first annual meeting of the Southeastern Association of Fire Chiefs. With the change of a few words, the talk would be as appropriate today as it was in 1928:

> Any person who is at all conversant with fire safety knows that at least eighty-five percent of [fires] could be prevented. It is the duty of the Fire Chief to assume leadership and point out the way for the protection of life and the conservation of property of our citizens.
>
> If the fire loss of the country is to be reduced, we must get away from the out-of-date methods of the old time red-shirt brigade. They thought the duties of the firemen were to sit around the engine houses waiting for an alarm of fire and then proceed to extinguish it as best they could; but the modern Fire Chief knows that he must be up and doing and prevent fires from starting, if he is to be successful in reducing the loss.

Chief Brosnan highlighted the reduction of fires in Albany in the previous five years. The following reduction occurred despite a 40 percent population increase: 1923 (160 fires); 1924 (159 fires); 1925 (125 fires); 1926 (110 fires); 1927 (97 fires).

Albany maintained a night patrol by uniformed firefighters. These individuals were detailed from fire companies and were required to look into every mercantile occupancy once each hour. In 1927, this patrol discovered more than 85 percent of mercantile fires, thereby keeping losses at a minimum.[1]

It must be recognized that a tremendous amount of progress has been made in fire safety and prevention. The popularity of smoke alarms and fire sprinklers is an example. Public fire and life safety education is an accepted program of most fire departments; however, thousands of citizens of all ages are still not fortunate enough to have the benefit of this important information.

Great strides have been made in hotel–motel fire safety. Federal legislation mandating that government employees on official travel stay in fire-safe properties is a major step forward. There are many other examples of fire safety improvements in publicly occupied buildings.

Figure 14.1 ◆ Live fire training is an essential part of fire extinguisher indoctrination by fire departments. *(Lukfin, Texas, Fire Department)*

Offsetting some of the progress in fire safety is the reduction of the staffing of fire prevention bureaus in many cities as an economy move. The widespread practice of utilizing fire service personnel for emergency medical duties has undoubtedly resulted in the saving of hundreds of lives, but it has forced some communities to eliminate fire company inspections and prefire planning surveys.

Fire service leaders recognize that the fire department alone cannot successfully impart the message of fire prevention and safety without strong community involvement. In many inner-city neighborhoods and low-income rural areas, individuals who are known and respected by the residents may be quite effective in generating an interest in fire prevention.

Fire prevention chiefs must be able to adjust their programs to meet new challenges such as the urban–wildland interface problem and the continuing expansion of the use of hazardous materials. Bureaus incapable of change will not survive.

◆ MEASUREMENT OF FIRE PREVENTION EFFECTIVENESS

Measuring the success of fire prevention efforts is subject to many variables.

As an example, a nationally circulated magazine attempted a comparative evaluation of fire departments through the use of per capita fire losses. Cities over the country were rated as to their fire departments' efficiency according to this criteria. The fire department protecting a bedroom community had a definite advantage over a community in which there was a good deal of heavy industry with little residential area.

Fire death totals have also been used as a means of measuring the effectiveness of fire prevention activities. Although such a comparison may have some validity, there are weaknesses in it as well.

Research has shown that socioeconomic factors have a bearing on fire safety measures and that a fair comparison of individual procedures in a community should take these into account. As an example, most cities experience a higher per capita fire death rate in low-income, row-house areas than in high-income, detached house areas.

Another measurement of fire prevention effectiveness is the relative position of the municipality from a fire insurance rating standpoint; however, the insurance protection classification does not directly reflect loss except in some major cities and counties.

Reports of achievement are also prepared at the national level. The U.S. Fire Administration convened a group of fire safety leaders in 1987 to review progress made in the field since issuance of *America Burning* in 1973. The report of this workshop, entitled *America Burning Revisited,* includes the following summary of achievement:

> The annual number of fire deaths has decreased by 23% from 1975 to 1985 (an average annual reduction of 1900 deaths). On a cumulative basis, this reduction means that an estimated 6,900 lives have been saved between 1975 and 1985. Fire fighter deaths also have been reduced significantly.

> Even greater reductions in fire deaths have been achieved within special categories. Clothing fire deaths have fallen by 73% over the period 1968–1983. In addition, children's clothing fire deaths have dropped by 90%.

> The number of fires reported to fire departments has decreased by approximately 20 percent over the period 1975–1985. The number of reported fires has been decreasing even though the population is increasing, which means that the number of fires is declining even on a per capita basis.[2]

Fire incidents are starting to show a decline as is per capita dollar loss, according to a 2001 publication of the National Fire Data Center of the U.S. Fire Administration. Civilian and firefighter fatality and injury rates are down as well. The publication gives credit to the following factors: smoke alarms, sprinklers, strengthened fire codes, improved construction techniques and materials, public fire safety education, firefighter equipment, and training. The statement concludes that "If we could understand the relative importance of these factors to lessening the fire problem, resources could be better targeted to have the most impact."[3]

Burn treatment has improved considerably in recent years. Over 100 centers provide burn care, with 25 of these offering full service burn treatment, research, and rehabilitation as compared to only 12 full service centers in 1972. Deaths, once the patient has entered such a center, have decreased from around 4,000 to 1,000 over 30 years.[4]

Another measure of success for fire departments today is the degree to which they practice customer service. The Phoenix, Arizona, department has been a pioneer in this respect. Customer service in the fire prevention field differs from that of responders to fires and medical emergencies. The customer to the fire prevention bureau may be a contractor, an architect, an arsonist, a fire code violator, or a citizen who is upset because the landlord has failed to replace a defective smoke alarm. The relationship and service concept vary in each contact. The old adage "The customer is always right" doesn't always fit in the fire prevention bureau. Of course, courtesy and professional conduct should always apply. An arrogant, overbearing attitude on the part of even one member of the staff can result in disaster for the bureau. The fire department scores very high in polls of government agencies in which citizens evaluate persons and organizations in which they have confidence.

The Fire and Emergency Services Self-Assessment Accreditation process sponsored by the Commission on Fire Accreditation International also provides a means for analyzing fire prevention activities. Fire departments meeting the provisions of this process are expected to have a fire prevention code in effect, with adequate staffing and enforcement as well as other support procedures. They are also expected to have an effective public fire safety education program directed to specific target groups and an adequate method for investigating fires. This program is spearheaded by the International Association of Fire Chiefs.

Some fire prevention analysts consider the percentage of a fire department's budget which is designated for fire safety work as an indicator of the community's interest in this phase of activity. Speakers at fire service conferences have bemoaned the average of 5 percent or less as reported by many career departments. When the potential value of fire prevention is considered, this is a very meager amount.

Brunswick, Georgia, a city of 26 square miles with a population of over 15,000, devotes 16 percent of its operating budget to fire prevention. The 37-member career department protects a seaport city that relies on tourism and industry as its economic base. Average annual fire losses are in the $100,000 or less range with few fire fatalities and injuries through the years. All fire department personnel engage in some form of fire prevention activity.[5]

An example of an assessment methodology as prepared by the Office of Fire Marshal of the Province of Ontario follows:

Inspections

1. Determine the level of compliance with fire safety legislation by the:
   - incidence of compliance within a specified time (determined by follow-up inspections)
   - incidence of continued compliance (determined at subsequent routine inspections)
   - incidence of continued compliance by properties not subject to routine inspection (determined by spot checks)
   - incidence of voluntary compliance by properties not subject to inspection
2. Provide an indication that the fire safety measures implemented in the inspection program reduced the impact of fire on evaluated incidents:
   - fire hazard elimination and reduction measures reduced the size and spread of fires;
   - containment measures limited the size and spread of the fires;
   - alarm and detection measures limited the spread of the fire and provided occupants with adequate warning to respond appropriately;
   - suppression measures limited the damage caused by fires;
   - fire safety planning information received by occupants limited fatalities, injuries, and damage caused by fires.
3. Provide an indication that the inspection program was responsible, in whole or in part, for fire loss reductions. This could be achieved by a thorough fire incident evaluation of all fires in properties subject to inspection.
4. Determine if fire incidents in the community may have been prevented as a result of the inspection program. This could be achieved by conducting surveys to seek information about the fire incidents.

Public Safety Education

1. Determine if there was an increase in fire safety knowledge as a result of the program by surveying target audiences before and after the implementation of the program. Establish a starting reference point and compare all subsequent results to it.

2. Determine from the fire incident evaluations that an increased knowledge of fire safety of occupants, as a result of the program, reduced the impact of evaluated fires.
3. Determine if the education program was responsible, in whole or in part, for fire loss reductions by conducting a thorough fire incident evaluation of all fires.
4. Determine if the fire incidents in the community may have been prevented as a result of the public safety education program. This could be achieved by conducting surveys to seek information about the fire incidents.

Fire Incident Evaluation
Determine whether sufficient information is being gathered through the fire incident evaluations to allow for a proper needs analysis and evaluation of the programs.[6]

The implementation of these provisions is having a major impact on fire fatalities in Ontario. A report issued in 2004 confirms this finding:

The Office of Fire Marshal (OFM) of Ontario reports a significant reduction in deaths due to fires since the provincial government enacted legislation in 1997 containing mandatory public education and fire prevention provisions.

In the six years since the ground-breaking law was passed, the fire death rate in Ontario dropped from 13.8 per million population in 1997 to 7.9 in 2002.

OFM is proactive in ensuring that municipalities meet their responsibilities under the Fire Protection and Prevention Act, 1997. Municipalities are monitored for compliance. If they meet a minimum acceptable model, as defined by the OFM, they are issued certificates of compliance.[7]

◆ ANNUAL REPORT

Many fire departments prepare an annual report to give community residents and the official governing body an accurate picture of the fire department's operation. The annual report should be simplified so it can be readily understood by members of the general public. The report should be attractive and easy to read. The use of photographs and charts will improve the appearance of the publication.

A number of communities do not prepare annual reports because of the cost involved. A tight budget should not preclude the issuance of a report. Even a photocopied report can be successful in advising the public of fire department activities. Arrangements should be made to make the annual report available to the public. It may be desirable for the report to be made in two forms—one a more complete report for use by public officials; the other, a summary report for distribution to citizens.

Norfolk, Virginia, a major port city with a population of almost a quarter of a million, uses its Fire Rescue Annual Report to give the citizens comprehensive information on the activities of the Fire Marshal's Office as well as other elements of the agency's activities. A recent report noted the passage of the previous year without a fire fatality, a major accomplishment for a city of this size. The addition of a fire safety house and advent of a procedure whereby civic neighborhood associations are given fire prevention programs geared to their own conditions are noted in the same report. Details of training classes attended by office personnel are also included, an area often overlooked in fire department reports.

The Norfolk report also includes a chart indicating inspections conducted during the previous year plus an enumeration of other tasks carried out. The activities of the

Arson/Explosive Investigators are also enumerated. These individuals operate under the Fire Marshal's Office. Closely related to fire prevention is the Environmental Crimes Task Force. These specially trained persons, also part of the Fire Rescue Department, inspect industrial facilities to reduce the possibility of accidental releases and minimize potential harm to citizens. They inspect likely premises and are empowered to issue permits and obtain warrants with subsequent fines. Their activities are also included in the annual report.[8]

◆ RECORDING FIRE SAFETY ACTIVITIES

The fire reporting system should also include information regarding activities of the fire prevention bureau, including demonstrations, lectures, other public education programs, plan reviews, and all other activities that contribute to the goal of reducing fire losses. Any public contact has a bearing on the official function of the organization and should be included.

The training division of the fire department should maintain complete records on time devoted to fire prevention training. The training division may call on the fire prevention bureau for assistance in such instruction.

A fire department record system should include information relating to all responses made by the department. These responses should be classified and enumerated in such a manner as to give an accurate picture of services performed by the fire department. Simplicity and uniformity must be major considerations. A separate report should be maintained on all inspections, prefire planning visits, and other activities carried out by the fire company.

FIRE PREVENTION INSPECTION RECORDS

Fire prevention bureau records should include a record of all inspections including each building subject to inspection by the department. Because the ownership and occupancy of a structure often change, problems may be encountered in attempting to maintain records of inspections under the names of owners or occupants. The owner or occupant may be listed on a cross-file system with the primary inspection record maintained by street address. The inspection record of a building should include the findings of each inspection and the disposition of requirements imposed as a result of these inspections. The file should also include a diagram of the structure, which might be used in the department's prefire planning program. Computers are very helpful in this process.

The file should include photographs of the building and of conditions found within it. Photographs and diagrams can be used to pinpoint the location of specific problem areas.

Copies of any orders issued in connection with the occupancy should be included in the file. Newspaper clippings regarding the occupancy can also be helpful. A news item may indicate policy changes by the operators of the structure and give other information that can be of great value in the event of a fire, such as changes in business conditions that might relate to the cause of a fire. Fire prevention bureau records with the exception of open arson cases, are considered public records in most states. Subrogation, arson cases, as well as other litigation in which details of fires must be recounted make the maintenance of adequate and complete records mandatory. It is necessary to be able to immediately recount inspections conducted and to have information available regarding fire prevention education programs carried out by personnel.

◆ RECORDING FIRE DEATHS AND INJURIES

Fire department records should include facts about injuries sustained in fires as well as about deaths incurred as a result of fires. Many communities do not have complete records because in most cases they record only deaths and injuries in which fire department response was involved. Medical reports are usually not public records.

A death or injury by fire in which the fire department was not called and the victim was taken directly to the hospital or mortician may be unknown to the fire department. This is often the case for burn injuries in which the victim is transported for medical treatment by a relative or neighbor and no fire department emergency medical or fire response is involved. Fire departments need to maintain close ties with hospital emergency rooms and the medical examiner or coroner's office.

Statistical reporting of deaths and injuries related to fire has another inherent obstacle. There are many borderline cases in which it is difficult to determine whether the death or injury should be attributed to fire. The question of assignment arises in traffic accidents in which the vehicle burns. A question as to whether the death or injury should be counted as a fire death or as a traffic death often arises.

According to the National Fire Protection Association, the risk of dying in an unintentional building fire has fallen from more than 20 deaths for every 200,000 people to just 3 deaths for every 200,000 (between 1896 and 1996). These fire fatality decreases can be attributed to many factors as described in earlier chapters.[9]

Approximately 17 percent of the nation's total fire deaths occur in vehicles. These incidents are not easily abated by fire service activities as normally carried out. Well over half of vehicular fire deaths occur as a result of a collision, an occurrence which relates to many factors: vehicle construction, driver ability, road configuration, and a myriad of other factors, none of which are under the control of fire prevention agencies.[10]

Certainly the elimination of fire fatalities is the major goal of public fire service prevention activities. Merely recording rote figures on fatalities is not enough. An analysis of fire death trends by cause can be quite helpful in determining proper avenues for fire safety education and enforcement activities.

◆ RECORDING LOSS STATISTICS

Widespread involvement of the fire service in the provision of emergency medical service has necessitated the development of data systems to record these responses. Many of these systems function at the state, as well as local, level. Motor vehicle injuries and fatalities, falls, cardiac arrests, drownings, and other non-fire-related incidents are recorded in connection with fire service response.

The availability of statistics on non-fire-related responses has been the driving force behind the expanding involvement of fire services in educational programs aimed at abating these injuries and fatalities. Fire departments may be joining with other community departments and organizations in disseminating information designed to lower the frequency of such incidents. Examples include programs promoting seat belt usage, CPR training, pool safety, and fall prevention.

The fire department record system should include information on fire losses in the community. Included should be data relating to causes of fires, buildings and occupancies in which fires occur, and information related to losses incurred in these fires.

Obtaining loss figures for all fires is a problem because there are so many variables in developing such information. Many departments use loss statistics developed by fire department personnel. Others use figures developed by insurance loss adjusters. In some cases, the departments count only losses not covered by insurance carriers. They believe that any dollar loss covered by insurance is not actually a fire loss and, therefore, does not count in their statistical report. Some fire departments classify losses as moderate, medium, high, or in other terms of a general nature. This "broad range" procedure makes it easy for personnel to complete their reports on the fireground; however, accuracy is not served by this practice. Personnel assigned to estimating losses need training in evaluation methods.

Records and reports are quite helpful in determining the value of fire protection devices such as fire sprinklers and smoke alarms. Statistics have authenticated the value of automatic sprinklers; however, smoke alarms are rapidly coming of age in this respect. Many communities are recording all smoke alarm operations in an effort to prove the value of the device. Statistics on performance coupled with examples of successful incidents in which smoke alarms and sprinklers saved lives can be of great assistance in promotional campaigns by fire departments.

◆ NATIONAL FIRE INCIDENT REPORTING SYSTEM

The National Fire Incident Reporting System is operated by the National Fire Data Center, a part of the U.S. Fire Administration. This system is designed to collect and analyze fire incident and casualty data on a nationwide basis. The standard *Uniform Coding for Fire Protection,* National Fire Protection Association No. 901, is used as the basis for the system.[11] Forty-nine states and the District of Columbia now report data. The U.S. Fire Administration is using the data for producing national estimates and for developing a database for answering information requests from government, industry, and the public. The data center is feeding national summaries back to the states for their use in evaluating and improving fire protection programs.

Data for the system, including fire incidents and casualty reports, are collected at the state level from local fire departments. Statewide summaries are prepared, as are feedbacks to local fire departments. Statewide data are sent to the system, which is the largest national annual database of fire incident information in the world.

It is imperative that fire incident data, whether gathered on a local, state, provincial, or national level, be of value to the agency submitting the initial report. Fire department personnel, whether career or volunteer, rapidly lose interest in supplying accurate information to a system that provides their department with few or no returns.

◆ USE OF COMPUTERS

Microcomputers are now widely used in fire prevention bureaus. The information age has provided the computer as a means of expediting decisions in fire safety code enforcement. The fire prevention chief is able to immediately review inspection criteria as well as results of code enforcement activities.

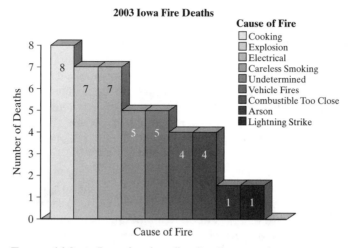

2003 Iowa Fire Deaths

Cause of Fire
☐ Cooking
☐ Explosion
☐ Electrical
☐ Careless Smoking
☐ Undetermined
☐ Vehicle Fires
☐ Combustible Too Close
☐ Arson
☐ Lightning Strike

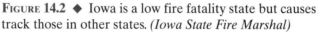

FIGURE 14.2 ◆ Iowa is a low fire fatality state but causes track those in other states. *(Iowa State Fire Marshal)*

The term *computer* is derived from the Latin verb *computare,* which translates as "to consider." A computer is a mechanical device capable of processing data. The term *microcomputer* refers to the smallest computer with full-function capabilities. Even one-person fire prevention bureaus are now taking advantage of this technology.

Computers reduce all information to simple electrical signals. This is accomplished regardless of the complexity of the information. A computer has a memory component, a control component, and an arithmetic component, as well as input and output capabilities. The memory component may be referred to as the storehouse for the computer. The actual physical portions of a computer are referred to as the hardware. Internal systems that enable the computer to carry out its functions adequately are known as operating systems. Programs that allow the user to perform specific applications on the computer are called software.

Several general software applications (programs) are available. The most popular is that of word processing, a process through which all types of documents—form letters, correspondence, and invoices—may be developed, revised, edited, and printed.

MARYLAND SMOKE ALARM PERFORMANCE

3 YEARS AVERAGE 2001–2003

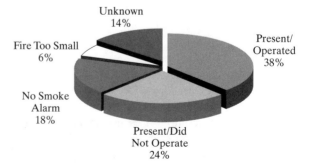

FIGURE 14.3 ◆ Of Maryland residences that experienced a fire, smoke alarms were present and operated in 38 percent. *(Maryland State Fire Marshal's Office, 2004)*

Fire safety survey reports are being prepared using word processing software in fire departments throughout the United States.

Another type of software provides for electronic spreadsheets. These may be used in much the same manner as the records kept by an accountant or bookkeeper. Budget totals and subtotals can be readily computed, as can hydraulic calculations for automatic fire extinguishing systems and water supplies for fire protection.

Database management programs are also available. These may be used to store and retrieve sizable amounts of similar data, such as records of inspections, hydrant flow test data, and fire extinguisher maintenance records.

Computers may be configured to transmit data from one to another using satellites or telephone lines. Fire service information is available through several national systems using this phenomenon. DSL, broad band, high-speed cable, and other innovations are enhancing the speed and ease of information transfer. In some cities, microcomputers in fire stations link the entire department through a telecommunications computer network. Routine correspondence, newsletters, messages regarding hydrant and street closings, duty reassignments, inspection assignments, and other routine messages are handled in this manner.

Graphics can be developed with the appropriate software. Graphics may be used in annual reports, budget presentations, and other projects where visualization is important.

Computers play a major role in fire service management. These devices have enabled fire service managers to immediately ascertain trends and directions that may be significant when planning for fire safety within a community. Likewise, they play a major role in fire protection research. A computer can give the researcher accurate information on probabilities that may be difficult to emulate in the laboratory.

In many fire departments, all firefighters are proficient in the use of microcomputers. Operational requirements demand that this degree of proficiency be reached if inspection programs, company response reports, company budget maintenance, and other essential activities are routinely processed by computer.

Computer-aided dispatch is becoming routine. This system usually includes methods of tracking the status of all fire fighting resources, occupancy data for hazardous and target locations within the city, and information on fireground operational requirements. Terminals for receipt of necessary data may be provided on fire apparatus and in command vehicles.

An example of the use of computers for fire code enforcement is the program through which fire departments may report inspections and fires. The city also records and prepares reports to property owners for inspections made by local fire stations. Each station is given a printout with information on the status of all fire inspections. The jurisdiction receives copies of the reports and is able to see the number and types of deficiencies within the community on receipt of each printout. This enables the fire prevention officer to keep track of inspections and to make appropriate personnel assignments from month to month. Many cities in the United States use a similar system for their fire inspection reporting and inspection frequency control. Other routine activities of the fire prevention bureau can be handled in a similar manner through the use of computers. Some track whether the person involved in a fire had received fire safety training. Handheld computers are also in wide use in fire services.

Computer applications in fire safety management, plan reviews, statistical packages, fire modeling programs, and the use of geographic information systems in determining necessary development planning information are other examples of the widespread use of this equipment. Fire modeling is being used in reconstruction of fires in connection with cause and origin investigations as well as in planning for fire protection.

The Occupancy Vulnerability Assessment Evaluation (called RHAVE for Risk, Hazard and Value Evaluation), a system developed by the U.S. Fire Administration in cooperation with the Commission on Fire Accreditation International, Inc., is an example of a computer-based system for analyzing and classifying occupancies under the purview of the using fire service agency. This system stores information on seven major factors relating to a give structure. Examples are provided for each factor.

1. PREMISES: Assessed valuation, area, revenue benefit, description
2. BUILDING CONSTRUCTION: Type of construction, access, size
3. LIFE SAFETY: Occupancy load, mobility, warning alarms, exiting system
4. RISK: Regulations in effect, human activity, loss experience
5. WATER DEMAND: Fire flow required and available, sprinkler protection
6. VALUE: Personal, family business, impact to community
7. SUMMARY: A compilation of the hazard involved based upon all factors above with resulting score of maximum, significant, moderate and low[12]

◆ APPROACHES TO THE FIRE PROBLEM

The following comments by New York City's fire commissioner's public information office provide an excellent summary of the approach of the nation's largest city to the fire problem, an approach that can well be emulated by communities large and small:

> The Department is engaged in a multifaceted approach to the fire problem that, although employed in the past, now is energized with improved management and communications capabilities. This coordinated effort includes virtually our entire department: operations, fire investigations, fire prevention and fire safety education.
>
> At fatal fires or those deemed suspicious by the Chief-in-Charge, we seek a determination of cause and origin, which is the responsibility of our Bureau of Fire Investigation. The probe conducted by Fire Marshals and their evaluations of causation allow us to be proactive in fire safety, prevention and enforcement efforts.
>
> When and where necessary, we mobilize our Fire Safety and Education Unit to target communities with an increase in the number of fires and/or fire-related fatalities. In collaboration with the Fire Safety Education Fund we have hired fire safety educators to distribute literature and give presentations to community boards, schools and various other community groups about fire prevention and safety. We also created a fire safety house at the Fire Museum and acquired two traveling fire safety trailers so that we can take the message of fire safety directly to schools and events where children are present.
>
> In 2002, the number of civilian fire fatalities (97) in New York City plummeted to its lowest level in 75 years. The last time civilian fire fatalities dipped below that level was in 1927.[13]

Plano, Texas, located north of Dallas, is one of the fastest growing cities in the United States. It has grown from 34,000 to over 238,000 population since 1973. Early in the 1980s, the city developed a master plan for fire protection that recognized expenditures for fire protection as including fire suppression and protection (the fire department); fire detection and suppression systems; fire and building codes; fire insurance premiums paid by businesses and residential occupants; and fire losses (direct and indirect). The city council recognized that the best fire protection system is the blend of all five elements, which results in the lowest bottom-line costs.

Since 1982, this blend, which includes a strong reliance on fire safety education, has resulted in savings to the city and to property owners and operators. A high percentage of commercial, educational, and public buildings, including fire stations, are fully protected by automatic sprinklers, thereby reducing water supply needs (needed fire flows). The city obtained a Class 1 grading from the Insurance Services Office, which has brought about a reduction in insurance rates.

The rapidity with which flashover conditions are reached as a result of more combustible contents in today's structures is an element that Plano is addressing with an aggressive fire prevention and fire safety program, which includes emphasis on installation of fire sprinklers. All fire department personnel are cross trained to deliver both fire and emergency medical services. All field personnel are trained and qualified to conduct fire prevention inspections and public safety educational activities.[14]

Prince George's County, Maryland, has required automatic sprinkler protection installation for all new one and two family dwellings constructed after December 31, 1991. Larger residential units also require this protection.[15]

David A. Lucht, founding director of Worcester Polytechnic Institute's Center for Firesafety Studies, prepared a list of six goals he feels the country should strive for in the next quarter century. This list was prepared in 2004, at the time of the 25th anniversary of the Center's establishment of the nation's first master's degree program in fire protection engineering. Over 300 master's and doctoral degrees in fire protection engineering have been granted since the program's inception.

Professor Lucht's suggested goals for reducing the nation's fire fatality rate follow:

Enforcement—Fire safety laws should be enforced with the same vigor as laws against speeding and drunk driving on the highways
Legislation—Laws should be modernized to give enforcement officials the same enforcement tools as are available to traffic cops
Public actions—Grass roots movements like Mothers Against Drunk Driving (MADD) can cause change in our culture tolerance of "fire traps"
Public education—Ten times more effort is needed nationally to educate the public about safety behaviors, especially for the most vulnerable groups like children and the elderly
Research funding—Fire research funding in the United States has declined 85 percent in the past 30 years. This trend needs to be reversed at the highest policy levels of government
Technology for fire services—Technological innovations are needed to help the nation's firefighters combat fires more safely and effectively[16]

The aggressive programs under way in Scottsdale, Arizona, Prince George's County, New York, and Plano exemplify the direction communities with a genuine interest in addressing the fire problem need to follow. Only through communitywide interest, spearheaded by fire service, can the nation's fire problems be seriously abated.

◆ **SUMMARY**

Records are essential for the proper administration of any fire prevention program. Accurate and complete information regarding fire prevention and fire control activities should be available if a thorough job is to be done. Analysis of statistical records can reveal specific problem areas toward which fire prevention education programs should be directed. Analysis of records can likewise point out the need for changes in fire prevention codes, modification of statutory requirements, and other changes in regulations.

Examples taken from records of the local fire department are far more meaningful than are reports of fire conditions taken from communities hundreds or thousands of miles away. Last, statutory requirements make it mandatory to maintain records of information necessary in subrogation cases, arson cases, and other legal proceedings. Court cases in which fire departments have been sued make it imperative to keep records of all fire inspection activities and fire prevention education programs.

Records and reports of fire suppression activities and fire prevention efforts are closely allied, and it is sometimes difficult to draw a line between the two. The types of reports and records the fire department should generally maintain are a complete record of fire department responses; statistical records of fire deaths, injuries, and fire losses; records of fire prevention and training activities; and reports for public information.

A complete record of all fire prevention inspections is all-important in the work of the fire prevention bureau to ensure proper follow-up of requirements imposed or changes necessary to conform to fire code regulations. This file should be kept up-to-date with any new information acquired.

From all these records, many fire departments compile an annual report that gives community residents, and the official governing body, and accurate picture of the fire department's operation and accomplishments.

Review Questions

1. A service added to the fire department that has greatly reduced fire company inspections is
   a. special rescue
   b. EMS
   c. hazardous materials
   d. terrorism
   e. rapid intervention

2. New challenges needing the attention of fire prevention officials include:
   a. fire watch
   b. water rescue
   c. urban–wildland interface problems
   d. urban search and rescue
   e. NFPA 1710

3. Most cities experience a higher per capita fire death rate in
   a. high-rise buildings
   b. garden apartments
   c. detached houses in high-income areas
   d. low-income housing
   e. hotels and motels

4. Between 1968 and 1983, children's clothing fire deaths dropped
   a. 50 percent
   b. 60 percent
   c. 70 percent
   d. 80 percent
   e. 90 percent

5. Deaths, once a patient has entered a full service burn treatment center, have decreased from 4,000 to _____ over 30 years.
   a. 1,000
   b. 1,500
   c. 2,000
   d. 2,500
   e. 3,000
6. A U.S. Fire Administration publication gives credit to what factor(s) for reducing civilian and firefighter fatalities?
   a. smoke alarms and sprinklers
   b. firefighter equipment and training
   c. strengthened fire codes
   d. public fire safety education
   e. all of the above
7. Inspection records should be maintained by address because
   a. owners and occupants change
   b. building construction types change
   c. use changes
   d. none of the above
8. Approximately _____ of the nation's total fire deaths occur in vehicles.
   a. 4 percent
   b. 8 percent
   c. 12 percent
   d. 17 percent
   e. 20 percent
9. The value of sprinklers and smoke alarms is enhanced by
   a. good records and reports
   b. manufacturing advertisements
   c. false alarms
   d. none of the above
10. Civilian fire fatalities in New York City in 2002 dropped to 97, its lowest in _____ years.
   a. 25
   b. 40
   c. 50
   d. 75
   e. 100

Answers

1. b
2. c
3. d
4. e
5. a
6. e
7. a
8. d
9. a
10. d

Notes

1. Southeastern Association of Fire Chiefs, *Proceedings of the First Annual Convention* (Atlanta, Ga., 1928), p. 18.
2. Federal Emergency Management Agency, U.S. Fire Administration, *America Burning Revisited* (Washington, D.C.: U.S. Government Printing Office, 1990), p. 141.
3. Federal Emergency Management Agency, U.S. Fire Administration, National Fire Data Center, *Fire in the United States, Twelfth Edition* (Arlington, Va.: TriData Corp., 2001), p. 27.
4. American Burn Association, *2000 Fact Sheet*, www.ameriburn.org.
5. Personal communication to the author from Chief Lee Stewart, Brunswick, Georgia, Fire Department, 2004.
6. Fire Prevention Effectiveness Model, Office of the Fire Marshal, Province of Ontario, 2003, www.gov.on.ca/OFM/model/prevent3.htm.
7. Peter Faulkner, "Ontario Regional News," *Canadian Fire Chief* (Ottawa, Ontario, Spring 2004), p. 18.
8. City of Norfolk, Virginia, *Annual Report of Fire Department* (2002).
9. James A Grisanzio, "What Do Americans Know About Fire Safety?" *NFPA Journal* (Quincy, Mass., National Fire

Protection Association, May/June, 1996), p. 79.

10. *Fire in the United States,* p. 9.
11. National Fire Protection Association, *Uniform Coding for Fire Protection,* NFPA 901 (Quincy, Mass.).
12. RHAVE (Risk, Hazard and Value Evaluation system); developed by U.S. Fire Administration in cooperation with Commission on Fire Accreditation International, Inc.
13. Personal communication to the author from Public Information Office, Fire Commissioner, Fire Department of New York, 2004.
14. Personal communication to the author from Chief William Peterson, Plano, Texas, Fire Department, 2004.
15. Section 4-154, Amendments to Prince George's County Building Code, 1988.
16. Worcester Polytechnic Institute, Press release, 2004.

Index

Engineering, role of, 23–34
Engineers, plan reviews and use of, 58–59
England, historical fire prevention in, 1–3
Environmental Protection Agency, 170
Ethical issues, inspection and, 90–91
Evans, Powell, 41
Evidence, preservation of, 192–93
Exit facilities, maintenance of, 136
Exit interview, 83–84
Exit signs, 135–36
Explosives, control of, 63–65

F

FAIR (Fair Access to Insurance
    Requirements) plans, 154
False alarms, 35
Fatality rates/statistics, 23
    international, 203–6
    recording, 242
Federal Aviation Administration, 166
Federal Bureau of Investigation (FBI),
    162, 183
Federal Communications Commission, 170
Federal Emergency Management Agency, 29,
    168, 232–33
Federal Highway Administration, 166
Federal Housing Administration, 166
Federal Railroad Administration, 167
Federal Trade Commission, 170
Fees for services, 121–22
Fire alarms:
    code enforcement and control of, 62–63
    early use of, 2, 4
Fire and Emergency Services Self-Assessment
    Accreditation, 239
Fire and life safety education programs, public:
    civic organizations, role of, 42–43
    clinics and seminars, 43
    combining programs, 48–49
    community events, 43–44
    contests, 49
    Fire Prevention Week, 51–52
    future needs in, 53
    home inspection program, 40–42
    hospital programs, 47
    media publicity, 50–51
    promotional aids and activities, 49–50
    review of successful programs, 52–53
    role of, 19–20, 21–23
    in schools, 44–47
    scope of, 38–40
    Scout groups, 47
    smoke alarm programs, 51
    steps for planning, 39

volunteer fire departments, 52
wildland, 48
Fire departments. *See also* Fire prevention
    bureaus:
    changes in, 20–21
    fees for services, 121–22
    fire prevention functions, 112
    inspection programs, 119–21
    personnel in, 122
    relationship with other municipal agencies,
        122–26
    relationship with state agencies, 122
    sources of conflict within, 117
    volunteer, 52
Fire drills:
    home, 130–31
    industrial, 133
    institutional, 132–33
    private fire brigades and, 139–40
    school, 131–32
Fire escapes, early use of, 4–5
Fire extinguishers:
    code enforcement and control of, 63
    inspection of, 82–83
    proper use of, 133
Fire Incidence Data Organization (FIDO), 148
Fire Investigation, Research, and Education
    (F.I.R.E.), 198
Fire investigators, use of, 191–92
Fire marshals, state, 171–74
Fire Marshals Association, 184
Fire prevention, historical:
    in England and Scotland, 1–3
    in North America, 3–6
Fire prevention, use of term, 14–15, 19
Fire prevention advisory committee, 118–19
Fire Prevention and Control Act (1974), 164
Fire prevention bureaus, 5
    communication with public, 115–16
    description of, in New York City, 114–15
    generalists versus specialists, 113–14
    one-member, 113
    purpose of, 112–13
    quarters for, 115
    working hours, 116–17
Fire prevention effectiveness:
    annual reports, use of, 240–41
    applications, 246–47
    computers, use of, 243–46
    fatalities and injuries, recording, 242
    inspection records, 241
    loss statistics, recording, 242–43
    measuring, 237–40
    National Fire Incident Reporting
        System, 243